Alban Jasper Conant

Foot Prints of Vanished Races in the Mississippi Valley

Alban Jasper Conant

Foot Prints of Vanished Races in the Mississippi Valley

ISBN/EAN: 9783337250584

Printed in Europe, USA, Canada, Australia, Japan

Cover: Foto ©berggeist007 / pixelio.de

More available books at **www.hansebooks.com**

THE BIG MOUND AT ST. LOUIS, 1869.

FOOT-PRINTS

—OF—

VANISHED RACES

—IN THE—

MISSISSIPPI VALLEY;

BEING AN ACCOUNT OF SOME OF THE MONUMENTS AND RELICS OF
PRE-HISTORIC RACES SCATTERED OVER ITS SURFACE, WITH
SUGGESTIONS AS TO THEIR ORIGIN AND USES.

By A. J. CONANT, A. M.,

Member of the St. Louis Academy of Science, and of the American Association
for the Advancement of Science

ST. LOUIS:
CHANCY R. BARNS
1879.

PREFACE.

The first organized effort for the study and preservation of the antiquities of this continent was the formation of THE AMERICAN ANTIQUARIAN SOCIETY, which was incorporated by Act of the Legislature of Massachusetts in October, 1812. A goodly number of distinguished scholars of that State were immediately enrolled as members, as well as many other learned men residing in other States, who were made corresponding members. In 1821 this Society published its first volume of "Transactions."

Although nearly all the accounts of the expeditions of the earlier explorers of the continent of North America contained notes of striking memorials, which were met with in their journeys, such as mounds, embankments, fortifications and the like, still, all such notices were too meagre and superficial to give them any special scientific value.

Through the labors of the American Antiquarian Society, the attention of the scientific world was first called to the countless, and often imposing memorials of pre-historic America, and their proper study began.

More than three-score years have now passed since the date of the organization referred to, during which time volume after volume has appeared upon the subject. Accounts of special surveys, by private individuals and State geologists, have multiplied every year. Scientific associations all over the land, as well as numerous Archæological Societies formed for this specific purpose, have given the subject unremitting and serious attention. The Smithsonian Institute has also expended large sums of money in

exploration and in publishing the accounts of the multiplied discoveries of observers employed by that noble institution. Nor
has the interest in the subject been confined to America. Some
of the wisest archæologists of Europe have written learnedly
upon the works and migrations of the ancient inhabitants of
both continents of America.

Notwithstanding all this labor and study, the great questions
continually repeated, which were suggested when our antiquities
were first noticed, still remain unanswered; namely: Who were
the authors of these works? What was their origin, and what
were the causes of their disappearance? Were they the red men
found in possession of the continent at the time of its discovery?
I am not aware that the opinion that the red men were the authors of
the most extensive works, though maintained by some scholars
of high repute, is held by any who have given them personal
and thorough examination. The earlier travellers who stumbled
upon them in the wilderness, or on the prairie, express their astonishment at their magnitude and the skill displayed in their
erection. Captain Carver, in the account of his travels in the
years 1766-'7-'8, describes what he was convinced was a military work, which he accidentally discovered upon the bank of
Lake Pepin. This was long before it was known that America
had any antiquities. Concerning it he says that "its form was
somewhat circular, and its flanks reached the river. Though
much defaced by time, every angle was distinguishable, and
appeared as regular and fashioned with as much military skill
as if planned by Vauban himself." Again: "I was able to
draw certain conclusions of its great antiquity." "How a work
of this kind could exist in a country that has hitherto (according
to generally-received opinion) been the seat of war to untutored
Indians alone, whose whole stock of military knowledge has only,
till within two centuries, amounted to drawing the bow, and

whose only breastwork, even at present, is the thicket, I know not."

His testimony is selected from that of a multitude of early writers, because he could not have been prejudiced by the preconceived opinions or notions of others, and also because he was a man of military training, being a captain in the British army, whose conclusions would not be mere guess-work. The judgment of Brackenridge, Atwater, William Wirt and many other distinguished men, is in perfect agreement upon this point, namely: that they could not have been built by the Indians, as we know them, nor any people in like condition.

The first important question to be decided—if, indeed, it can be—is the origin of the North American Indians. Was their original home in Asia? and did they, as many believe, make their way to this continent across Behring's Straits, or the Aleutian Islands? In the solution of this question, the student of Archæology bespeaks the aid of the Philologist.

Archæology, Geology and Paleontology have been called three sister sciences. Philology must be added to the sisterhood, at least as far as American pre-historic times are concerned, as it may be that the lamp of this younger sister may light the footsteps of the elder to the final results of ethnological investigation. Should there be discovered some radical affinity between a few of the forty stock languages of the red men of North America and their physically-related brethren of Northwestern Asia, the question, it would seem, would be settled.

No one who has not previously examined the antiquities of the vanished races who once dwelt upon this continent, can have any adequate idea of their magnitude and extent; and he who has seen them as they lie thickly scattered throughout the fertile valleys of the Western States, is surprised at the evidence they

present of a prodigious population which once swarmed in this
wide domain.

The Cahokia group of mounds in the American Bottom, six
miles from St. Louis, may aid in illustrating the statement.
There are in this group, beside smaller ones, at least a dozen in
close propinquity; any one of which, if standing alone, would be
a striking feature of the landscape. Just west of Monk's Mound,
the largest of the group, is one which is perhaps fifteen to twenty
feet in height. On the flat surface of the top stands a good-sized
farm-house, with necessary out-buildings and conveniences, and
plenty of room for a variety of fruit trees, and an ample vegetable garden. This is not the only one so occupied, and there are
near this one several others much larger and more conspicuous;
while all of them look diminutive, and may be compared to ant-
hills beside that King of Mounds which stands near the northern
center of the group. Directly north of this great work are satisfactory evidences of a large artificial lake, the dirt taken from
which was presumably used in building the mounds. The form
and size of the lake can best be made out just after the Cahokia
creek, which flows through it, has been swollen by heavy rains.

Having his home in the center of the great valley where these
works abound, and looking upon some of them almost every day
of his life, the writer has been impelled to note the facts which
came under his own observation, and to venture, concerning them,
a few conclusions. The conclusions may be valueless; but if he
who causes two blades of grass to grow where but one grew
before, is a benefactor, truly, he who contributes a new fact to
the sum of human knowledge, should be considered in the same
light: for the after times alone can reveal the value of the con-
tribution, and that which is least in all earthly kingdoms sometimes becomes the greatest. In the desire to do what he could
to advance the study of a science so deeply interesting to himself,

the writer commits these chapters to the public, making no claim to originality beyond the accounts of his own explorations of the mounds and caves therein contained.*

It should be noticed that this memoir has already appeared in the voluminous work, entitled, "The Commonwealth of Missouri," and as there has been, ever since the appearance of that work, much inquiry from various quarters, particularly outside of Missouri, for the Archaeological portion, and by those who had no special interest in the local history of our State, and therefore did not care to incur the expense of purchasing so costly a work as the Commonwealth, it has seemed best, without further delay, to reprint the work from the original plates, as it first appeared in the history referred to. This statement will explain the frequent allusions which will be noticed, to that work.

A. J. C.

* It may not be amiss to state, also, that an additional interest may attach to the subject-matter of this work, because it is the result of explorations in an entirely new field, concerning which nothing has before been published. The account of every new "find" in the Archaeological field always elicits the attention of scientists all over the world. This is illustrated by the fact that a short paper upon the mounds in Southeast Missouri, read by the author of this book at a meeting of the St. Louis Academy of Science, and published in its transactions, has been translated and republished in France, Germany, Austria and Denmark.

TABLE OF CONTENTS.

CHAPTER I.

CHAPTER VII.

CHAPTER VIII.

CHAPTER IX.

CHAPTER X.

CHAPTER XI.

CHAPTER XII.

Man in the Age of the Mammoth and Great Bear.

CHAPTER I.

TRACES OF VANISHED PEOPLES. — THEIR WORLD-WIDE DIFFUSION. — RUSSIAN EARTH-WORKS. — EGYPTIAN MONUMENTS ANCIENT AT THE DATE OF OLDEST RECORDS. — A TROY STILL OLDER THAN THE ANCIENT TROY OF HOMER.

IN all lands, whenever in the ages past the climate has been such as to render it possible for man to subsist, the earth is found thickly planted with the graves of vanished peoples. Countless generations have come and gone, and left no record of their lives and work, save what is to be found in the few surviving monuments they have erected, or the rude implements and fragmentary remains of their industry, which descended with them to the tomb. The great ocean of humanity, with the energy of its ceaseless flow, has oft-times, no doubt, obliterated the traces of former generations, save here and there a foot-print in the solid rock, or an empty shell which has been left upon the shores of time. We of to-day build, sow and reap, buy and sell, and thus repeat, over and over again, the great drama of life, above the sepulchers of departed millions, long since forgotten. How often the long eons have finished their cycles and the new began—who can compute, or from whence shall the data be drawn upon which such computation may be based? The sacred records furnish no system of the chronology of the race, nor standing-ground upon which a trustworthy one can be constructed. The wisest who have essayed the task, from such sources, differ in their estimates more than five thousand years. The devout believer in Revelation, therefore, need feel no apprehensions for the foundations of his faith if it shall be proven even that man has been an inhabitant of the earth for a hundred thousand years or more.

All that can be gained from history, sacred and profane, supplemented with the hieroglyphic annals of Egypt and the inscribed bricks and cylinders of Assyria, carries us back only about forty-four centuries. Suddenly we come then to the border-land of legendary myths and extravagant traditions. The thick darkness which enshrouds all beyond, no one, a hundred years ago, thought possible to penetrate or dispel. But within the last fifty years a new science has been added to the varied departments of human knowledge and research—the science of Archæology, pure and proper,—and thousands to-day, including many of the best minds in the most enlightened lands, are devoting to it their serious and earnest labors.

The field of exploration is the wide world, whose continents are all equally rich in the monuments of the forgotten past. From the widely-separated quarters of this great field, the laborers gather from time to time, bringing the results of their work. All these combined are throwing their focal light upon the great questions of the origin and antiquity of the various races of mankind—their peculiar customs and mode of life—investing them with an interest never before awakened, which increases more and more, as the promise brightens of their satisfactory elucidation. The number of the monuments of which we speak, upon our own continent, is legion upon legion. From Nova Scotia to the southern coast of Florida—from Behring's Strait to Mexico and Peru—from the Atlantic to the Pacific—are to be found the sites of ancient cities, or the former seats of a dense population. Europe, as every one knows, is full of them. Not only on the surface of the soil, but far down in the gravels of the drift, are found the remains of man in companionship with the bones of huge mammals, who were buried there, it would seem, long before the "British Channel was scooped out." In Russia, from its western border to the Pacific, from its southernmost limit far north into the inhospitable regions of Siberia, earthworks are found giving evidences of long occupancy, and doubtless a forced migration to the North. There, in the sepulchers of the dead, they deposited the gold and silver ornaments and other treasures of the departed, in which relics the more recent inhabitants have driven a thriving trade. The great steppes of Asia abound with sepulchral mounds. Nor are the deserts of Africa without their witness to the existence of former generations. Her remorseless sands are the tomb of many an ancient city.

Egypt, the oldest nation which has preserved a written history, has also her pre-historic remains. Before the name of Athens was pronounced, or Greece was born—when Italy was peopled with savage tribes as wild and barbarous as the red men of America,—Egypt was far advanced in the higher branches of knowledge, the sciences and the nobler arts. Her priests even then dwelt in the palaces of the kings, and issued their mandates, with his, from the throne. Those palaces were colleges of learning, while the priests were the professors, who not only ministered in matters of religion and worship, but devoted themselves to the higher education of the young as well.

To-day, as the explorer removes the stones from her ancient structures, he finds here and there one, whose inner surface is carved with curious devices and inscriptions, showing that it once had a place in older and demolished edifices. She had then her libraries also, in which the knowledge of her sages was preserved. Tombs of the librarians have been discovered, dating back at least five hundred years before Homer sang in the cities of Greece, and inscribed "To the chief of books."

Long since, the line of the Pharaohs became extinct, and no prince or king—so the prophet said—shall ever sit on her throne again or sway the scepter over the land of the Nile. How old she seems! And yet old Egypt was of yesterday, compared with the men of the drift, the reindeer period, or the pre-glacial times of Scandinavia, Scotland, France, England and the Pyrenees.

These everywhere ancient monuments of which we speak, men have been wont to regard with unquestioning curiosity, or at most to pass by with a conjecture only, as Homer did, who speaks of the ancient mounds, concerning which, in his day even, there was no history or tradition, and who imagined they might be the tombs of ancient heroes. Job makes more than one allusion to the monuments and "solitary mansions of the dead," which awakened the curiosity of the caravans and travelers of Teman, as they passed along the great thoroughfares of commerce. The wild songs of the most ancient bards are no longer poetic myths, the creations of a fervid imagination; but their inspiration was drawn from events which actually transpired. "With truth their souls were fired." The poets were the nations' historians as well. Troy, with her strange story, is no longer a doubtful city. Dr. Schlieman has found her ancient site and discovered enough, among her long-entombed memorials, to authenticate her history; and we may write once more "*Ilium est*" for "*Ilium fuit.*"

And what is most surprising of all, far down beneath the level of the ground once trod by the heroes whose names Homer has given to immortality, the explorer has found the ruins of another city—and he thinks still another below it—concerning which the poet seems to have heard no tradition. Among those deposits of an age so remote, were articles of stone and bronze and precious metals, skillfully wrought, giving evidence of the existence of a people whose knowledge, attainments, and social condition were far in advance of those of the more ancient periods of stone and bronze—a civilization which could only have been realized by the slow growth of centuries.

But not alone upon that glorious land, made immortal by the fiery energy of Homer's matchless songs, has a resurrection morning dawned, nor Egypt and Assyria with their hieroglyphic annals, hoary with age; but other lands, unknown in classic story, and the islands of the sea, are giving up their long-forgotten dead. The explorers of to-day are breaking down the hitherto impassable barriers of the remoter ages of antiquity; here and there we catch glimpses of the life and customs, and hold converse with the tribes and peoples of pre-historic times. The fast-accumulating records which have been gathered during the last twenty-five years are continually enriching the libraries of every civilized nation, and he who would master them all will soon find life too short to do more than acquaint himself with the grand results of the multiplied discoveries. The chief difficulty then, it will be perceived, in the way of the present task, is one of condensation, or in other words, how to select from such vast material only those facts and observations which are necessary for the proper treatment of the subject we are about to consider.

On account of the limits prescribed for the archæological chapters of this work compelling all possible economy of space, and also for the sake of continuity, instead of burdening them with frequent references to the authors consulted, I desire in the outset to make all due acknowledgment of my indebtedness to those valued records of the labors of the noble army of abler men who have preceded me in like investigations in this department of knowledge. Chief among those which have been freely consulted, are the writings of GARCILLASSO DE LA VEGA, PROF. REFINESQUE, DANIEL WILLSON, LL.D., ALEXANDER W. BRADFORD,

J. W. Foster, Edward L. Clark, Wm. Pidgeon, Prof. G. C. Swallow, Sir John Lubbock, M. L. Figuier, M. Marlot, John Evans, Lewis C. Beck, H. M. Brackenridge, James Adair, and others. Also, an article upon the Archæology of Missouri, contributed by myself to the last volume of Transactions of the St. Louis Academy of Science.

CHAPTER II.

Methods of the Archæologist. — The Shell-heaps of the Baltic. — The Buried Forests of Denmark. — The Sisterhood of Science. — The Five Geological Periods. — The Ages of Stone and Bronze. — Iron in common use three thousand years ago.

As before remarked, in almost every land upon the surface of the globe, are to be found countless monuments and memorials of vanished races; sometimes structures of imposing magnitude, but oftener implements of war and the chase, of domestic use and personal adornment. From such remains, more or less rude and defaced, it has been found possible to reconstruct a pre-historic history of man's life in the most remote ages of his existence; and by their careful study we are able to scrutinize his manner of life; to look in upon his domestic scenes; to witness his ceaseless struggles for existence—his mode of burial; and to learn something of his notions of another life. Only one important thing is forever lost—his language. For "we can never hear him speak." Yet the history we may recover is as true and touching as any which the poets sing. Nor need all this be thought incredible, for these results are obtained by the simple processes of reasoning and induction which we apply to the affairs of every-day life. When the traveler upon our western plains stumbles by chance upon the ashes and debris of a former habitation, if he finds there the fragments of a hoe and sickle, he at once infers that the former occupant was a tiller of the soil; should his eye light upon a cast-off shoe of infantile proportions, he naturally concludes that once it was the home of childhood. In addition to this, should he discover charred bits of bread and other articles of food, carbonized grain and fruits, along with culinary articles, showing the action of fire, these facts would show what crops were grown, the kind of food upon which the family subsisted, and also that the dwelling was destroyed by fire. The presence of the fragments of a crucifix would point to the religious belief of the former occupant.

Such is the method of the archæologist. When he examines the huge heaps of shells along the shores of the numerous arms of the Baltic sea, composed of individuals of large size, select and full-grown, of several species, commingled with rude implements of stone and bone, with also the bones of the codfish, and compares them with the diminutive specimens he is able to procure from the same waters now, it is an inference most reasonable, that when these heaps were piled up around the miserable huts of the ancient fishermen, the waters of the Baltic were not so fresh as now. The presence of the bones of the codfish gives some evidence of skill in navigation, for they must be caught in the open sea. When the peat-bogs of this same country are examined, they present a record reaching far back of the historic period. These depressions in the natural surface of the earth—sometimes to the depth of thirty feet, disclose three distinct periods of arborescent vegetation. At the bottom are the stately trunks of the pine tree; above these the oak, which once grew upon the sides of the pits, and when their full maturity was reached, fell inward. The oak was succeeded by the beech and birch which now flourish—and have flourished during all the period of history—throughout the land. The pine and oak have never been known during the historic period in the native forests of Denmark. In these bogs, beneath the layers of pine, are found the rude implements of the ancient inhabitants. Man lived, then, when the pine forests were in their glory, and at that time also piled up the shell heaps along the shore; for in these are found in great abundance the bones of a bird whose food is derived from the pine.

Again: when the student of Archæology discovers — as is frequently the case—the bones of extinct mammals, in situ, each bone lying by its fellow in its relative position as when in life, he knows there can have been no disturbance of the remains since the death of the animal. If he finds also, in companionship with them, the relics of man's industry, he believes that these mammals and man were contemporaneous. Should he find, further, huge bones split longitudinally, and showing marks and scratches of flint knives, which could only have been made while the bones were soft, he naturally concludes that man hunted these animals for food and split the bones to obtain the marrow. But the generalizations of the archæologist are not based upon the study of such relics alone. Geology,

Paleontology and Archæology go hand in hand, and have well been called "three sister sciences." Each of these three related departments of human knowledge is throwing its focal light with increasing luster upon the great question of man's first appearance upon the earth. By the light of their combined disclosures, the steps of our groping feet are illumined as we travel slowly along the pathway which leads us irresistibly to the night of the unknown ages, "and the mind recoils dismayed when it undertakes the computation of the thousands of years which have elapsed since the creation of man."

The five geological periods into which the crust of the earth has been divided, are commonly named in the relative order of their age: the primitive rocks, the transition rocks, the secondary rocks, the tertiary rocks, and quaternary rocks. All of these are anterior to the present geological period. The long succession of animals and plants peculiar to each, is found generally to have died out during the time of its continuance. Judging from the present order of things, each period must

A Solitary Cave Dweller.

have been of long duration; for the animals and plants with which we are familiar show scarcely any alteration since their first appearance, though they have existed for thousands of years. Now it is considered certain by the best informed, that man existed in Europe at the commencement of the quaternary period.

We are not left in doubt as to the climatic conditions of that country in those remote times, which must have been similar to the polar regions of the North to-day. There was no Iceland, Scotland, or Scandinavia then. The whole continent was shrouded in a winding sheet of snow. Her now beautiful valleys were the bottom of the sea. Enormous ice-fields stretched away from mountain to mountain, and only the highest elevations of the Pyrenees and Apennines were visible above the vast expanse of eternal snow and ice. Yet there, during that awful winter, for

which there was no promise of a coming spring, man and cotemporaneous
animals contrived to exist.   But what a life! To us, it would seem
utterly hopeless and dreary; but for its maintenance he found abundant
employment for all his activities, in providing means for his daily
sustenance, and in his contests with the wild beasts around him for the
possession of the shelters of the caves and overhanging rocks.   How
long this period continued we cannot know; but the centuries rolled
on, and slowly the glacial period comes to an end—the ice-fields melt
away, the glaciers retreat to the north, and the submerged continent
arises from the ocean.   The sunshine and the genial air of a new spring
morning dissipate the tears from the face of Nature, and she hastens to
put on her robes of green.   With this dawn of another life a new

The Elephas Primigeneus.

generation of animals now makes its appearance on the earth, and very
different too, from those which perished during the glacial period.
Among them the huge mammoth (*Elephas primigeneus*) with his
woolly covering and lion-like, shaggy mane; the Siberian Rhinoceros
(*Rhinoceros tichorinus*, with curious horns) and clothing of fur, so soft
and warm; several species of the Hippopotamus; the Cave Bear, of
prodigious size, (*Ursus spelæus*); the Cave Lion (*Felis spelea*); various
kinds of Hyenas, the Bison, the Urus, (*Bos primigenus*), and the
gigantic Irish Elk, with enormous wide-spreading antlers, and many
others which need not now be mentioned.

These huge monsters rapidly multiply and roam in countless multitudes over the continent, as do the buffaloes of our western wilds to-day. Hundreds and thousands gather together in their favorite resorts and from some cause unknown they perish. How man could successfully contend with such formidable adversaries with the rude implements he was able to construct by his infantile skill, is surprising; but his necessities compelled him to be victorious. Nor was he then destitute of æsthetic taste; for at his leisure he carved in stone or bone the outlines of the beasts he had slain in the chase.

At length the long summer ends, and another fearful winter begins. Again the cold is intense; the glaciers advance through the valleys toward the south. The floods increase, the caves are submerged, and man seeks a home again in the mountain ranges. The valleys are filled with alluvium for hundreds of feet up the mountain sides. The centuries roll on — how long, no one can tell, — and again another subsidence of the floods, or uprising of the continent, takes place, and the glaciers once more recede to the north. Slowly the mountain tops are lifted toward the sky, and the earth is clad again in green.

Man now returns to the former abodes of his ancestors. But what a change has taken place! Many of the mighty mammals his forefathers hunted on the plains are seen no more. A few solitary individuals linger on, but soon he witnesses "the extinction and disappearance from the face of the earth of an entire fauna of the larger animals."

From this period the Reindeer epoch, — known also as the period of migrated animals — begins. A new civilization dawns. Polished implements of stone and bone take the place of rude chips and splinters of silex. Pottery is manufactured and ornamented with curious devices; and all that man does displays the awakening exercise of his sense for beauty. From this time the race proceeds with slow but steady advancement. How long the Neolithic, or polished stone period lasted, we have no means of judging, nor when men learned to smelt the more yielding ores, and to make bronze by the alloy of copper with tin. But when that great discovery was made by which he supplied himself with a material so much better fitted by its superior hardness to copper for cutting implements and other uses, he entered that pathway, which ends only in all the glorious possibilities of the future. With this discovery, the age of Bronze was ushered in. Speedily its use spread over the greater part of Europe. With the age of bronze the arts and sciences may be said to have had their birth. Of the time of its continuance, which seems to have been long, we know but little more than we do of

the age of stone. But at length it seems to have been brought to a
sudden termination by that mightiest physical event in the history of
the development of mankind—the discovery of Iron. As to the time
when this great transition took place, history is silent; for it was long
before history began. The poems of Homer and Hesiod prove that
iron was known and in use at least three thousand years ago.

CHAPTER III.

No "Age of Bronze" in America.—Traditions Regarding the Mounds.—Tuscarora
Chronology.—The Animal Mounds of the Upper Mississippi Region.—Ancient
Fish Traps.—Burial, Sacrificial and Historical Mounds.

The facts, and the conclusions they suggest, presented in the foregoing chapter, are gathered mostly from the continent of Europe.
Each of the great geographical divisions of the globe seems to possess
an archæological record more or less peculiar to itself. Our own
continent has had no age of bronze. At the time of its discovery,
however, implements of copper, beaten out usually, but sometimes
smelted and cast in a mold, from the native ore, were to some extent
taking the place of those of stone and bone. And although the copper
regions of Lake Superior, for the distance of more than one hundred and
fifty miles along it southern shore, give evidence of long-continued
mining operations upon a stupendous scale, still we must believe that this
metal was too costly to be to any great extent the property of the masses;
while, even in our own times the remnants of some savage tribes may be
found who still point their spears and arrows with stone. The presence
of the relics of such material therefore, it hardly need be said, is of no
value in questions of antiquity, only so far as they are found in companionship with the remains of extinct animals, or their age is demonstrated
by geological or some other irrefragable proofs.

But now, leaving all other facts and considerations bearing upon the
general subject of archæology, which might be interesting and appropriate
in this connection, it is proper to proceed to the examination of the monuments of our own land, among which those found in Missouri are
peculiarly instructive, not only as forming no inconspicuous part of the
one great whole, and calculated to shed much light upon the question of

the homogeneity of the vast population which once swarmed upon this continent, but also—if not their origin—at least the direction of their disappearance.

In view of the magnitude of the subject, the ethnological questions involved, and the evident relation of these remains to all which are found in both North and South America, it has seemed to me impossible to examine them in the most profitable manner, if our examination shall be circumscribed by the imaginary boundaries of the State. For the reason mentioned, I have also presented, as briefly as possible, the preceding statement of the results achieved by the labors of the archæologists of Europe. I will now proceed to speak of some of the more important monuments of this country, with such description as suits my present purpose.

The statement has been often repeated by writers upon this subject, that the Indians have no traditions concerning the authors or the design of these monuments. This is undoubtedly true as far as the degenerate remnants of the tribes of the present day are concerned. But when the country was first discovered, and long after, here and there a solitary individual was found who claimed to be a prophet, and to have descended from a long priestly line, and also from a race superior to the Indians by whom their forefathers had been conquered and enslaved. Concerning the traditions handed down from father to son, they were very reticent, except under peculiar circumstances and with those who gained their highest confidence and esteem. The sacred treasures of their history, of which they were the preservers and guardians, were not for the common masses of their own people ; much less would they communicate them to strangers and foes. And when, as it sometimes happened, their frigid reserve would be conquered, and a narration of their legendary history elicited, it was considered more wild and untrustworthy than the long lists of Manetho and Berosus, of Egyptian and Assyrian dynasties, and not worth preserving. From this cause many valuable facts have been irrecoverably lost. A few only have escaped oblivion, of which the briefest possible mention can now be made.

The traditions of the Wyandot Indians, according to the account of Mr. Wm. Walker, for some time Indian Agent for the Government, published in 1823, are not devoid of interest. They were in substance as follows :

Many centuries ago, the inhabitants of America, who were the authors of the great works in the Mississippi Valley, were driven to the south

by an army of savage warriors from the North. After many hundred
years, a messenger returned from the exiled tribes, with the alarming
news, that a terrible beast had landed on their shores, who was carrying
desolation wherever he went, with thunder and fire. Nothing could stay
his progess, and no doubt he would travel all over the land in his fury.

It is conjectured that this beast of thunder and fire referred to the
Spanish invasion of Mexico. The Tuscaroras, according to the account
published by Mr. David Cusick in 1827—quoted by Prof. Rafinesque—
had a well-arranged system of chronology, dating back nearly three
thousand years. Their traditions locate their original home north of
the great lakes. In process of time, some of their people migrated to
the river Kanawag (the St. Lawrence). After many years, a foreign
people came by the sea and settled south of the lakes. Then follow
long accounts of wars, and fierce invasions by nations from the north,
led by confederate kings and a renowned hero named Yatatan. Many
years again elapse, and the king of the confederacy pays a visit to a mighty
potentate whose seat of empire is called the Golden City, situated south
of the lakes; and so on, down to the year 1143, when the traditions
end. In these records appear accounts of wars with various tribes,
given with great particularity; migrations southward and west to the
Mississippi, (called Ouauweoka); the names of the ruling monarchs,
and the order of their succession. There appear to have been several
dynasties of longer or shorter duration. Thus, the name Tarenyawagon
is borne by three successive monarchs, and Atotaro is continued to the
ninth.

Only a few items are here given, to indicate their character. No one
can examine these traditions without being convinced that they have
some great historic facts for their basis, however incredulous he may be
as to the correctness of their dates, or their pretentions to so high
antiquity. The limits prescribed for this essay admit of but one more
notice of traditions in this connection.

A class of works, frequently noticed by explorers, is found on the
upper Mississippi, chiefly in Wisconsin,—a few in Ohio, Indiana, Illinois
and Iowa—known as animal mounds, on account of their striking
resemblance to the forms of various animals, such as the Buffalo, Bear,
Elk, and the like, and some to the human form. These works have
elicited much discussion and conjecture as to their origin and purpose, in
which no two writers agree. Some of them are of gigantic proportions,
and cannot be ascribed to the present race of Indians, for the same
reason which precludes the idea that they were the authors of the
stupendous works of the more southern States.

The traditions relating to these animal mounds are very minute, full and interesting, and were first published in 1853, by Mr. Wm. Pidgeon, who spent several years in the examination of the various monuments in Virginia, the Valley of the Mississippi and South America as well. He tells us that he began these researches from motives of personal interest merely, and continued them for several years, without any design of publishing the results of his observations.

During his travels in the regions of the Upper Mississippi, he met a stranger among the red men, of dignified and venerable appearance, who had no fixed abiding place, but wandered from tribe to tribe, always welcomed and venerated wherever he went; who claimed to have descended from a

Fig. 1.—Mastodon Mound.

long line of ancient prophets, he the last of the line and the last of his race. He was then nearly ninety years of age. The Indians called him "the Mocking Bird," because he could speak fluently five different languages. By kindness, his confidence and friendship were won, and his companionship secured during the journey of exploration. He seemed perfectly familiar with all the most important works, from the Ohio to the extreme north and the far west,—could draw their outlines from memory, and supply any defect in the drawings of others; and could generally give a ready and lucid account of their authors and the purposes for which they were constructed. Unlike many who have written upon the pre-historic people of America, the author seems to have had no pet theory to maintain—as that they were the ten lost tribes of Israel, and the like,—but to have been a thoroughly conscientious and careful observer, faithfully noting what he saw and heard.

From the seventy engravings—and accompanying descriptions—with which the work of Mr. Pidgeon is illustrated, I select two or three, and leave the reader to judge whether these traditions are reasonable and trustworthy or not.

Many years ago, in the bed of Paint Creek, in Ross County, Ohio, several deep cavities or wells were discovered, which gave rise to much speculation as to their origin and purpose. I believe they have since been found in many other localities. Mr. Pidgeon states that he discovered four similar ones in the bed of a small tributary of the St. Peters river, varying in depth from eight to twelve feet, from five to six feet in

diameter at the bottom and from three to five feet at the top. These
excavations were made in the soft slate rock which formed the bed of
the stream.

To the level top, or rim of the well, a thin flat rock was fitted, with a
round or square hole in the center, about twelve inches in diameter.
This opening could be closed at will, by a stone stopper perforated with
small holes. A short distance below the wells he found one of these
stoppers which fitted neatly the larger capstone of one of the wells. At
the time of their discovery the depth of the stream which flowed over
them was ten inches. Mocking-Bird informed him that these were fish
traps, and that many such could be found in other streams, were they not
so filled with mud and stones as to escape observation; and also that
they were constructed and used anciently for the purpose of securing a
supply of fish for the winter. Large quantities of bait being deposited
in them in the fall, the fish would gather there in great numbers, when
the stopper would be placed over the mouth, which prevented their
escape, and then they could be taken out with a small net as desired.
While it is no doubt true that the mound-builders were an agricultural
people, it is quite reasonable to suppose, from the fact that their most
extensive works are found upon the shores of lakes and banks of rivers,
that fish formed no inconsiderable item of their bill of fare.[1]

As before stated, the historian of these traditions, after the death of
Mocking-Bird, proceeded to investigate by careful excavation those
earthworks of which he had previously made only a superficial survey,
especially those concerning which he had received traditions. The first
group thus explored which I notice is represented in Fig. 2. It is
described as being located at the junction of Straddle Creek and Plumb
river, in Carroll County, Illinois. It is composed of conical mounds,
rings and semi-circles, with diameters varying from twelve to twenty-five
feet. The rings are about two feet high, and seem to have been formed
by throwing up the earth from within, leaving the interior in the form
of a basin.

The traditions concerning these works are in substance that they were

---

[1] Some writers have discredited the idea of the artificial origin of these wells or fish-
traps, attributing their formation to the disintegration of the rocks in which they occur,
owing to the unequal hardness of the strata of which they are composed, etc. But it
would seem that vastly more credulity is required to believe that the ordinary operation
of nature in various parts of the country would produce such cavities, from eight to
twelve feet in depth, with nice fitting covers, perforated at the center, than that they
are the workmanship of intelligent beings for some special purpose.

constructed by a people who were accustomed to burn their dead, and
were only partially occupied. Each family formed a circle sacred to its
own use. When a member died, the body was placed in the family circle
and burned to ashes; a thin covering of earth was then sprinkled over
the whole. This process was repeated as often as a death occurred, until
the inclosure was filled. The ring. was then raised about two feet and
again was ready for further use. As each additional elevation would of
necessity be less in diameter than the preceding, in the end a conical
mound would be the result. The darkest spots in the engraving
represent those which are finished; the rings, those in various stages
of occupancy; and the semi-circles those which were only begun. Similar

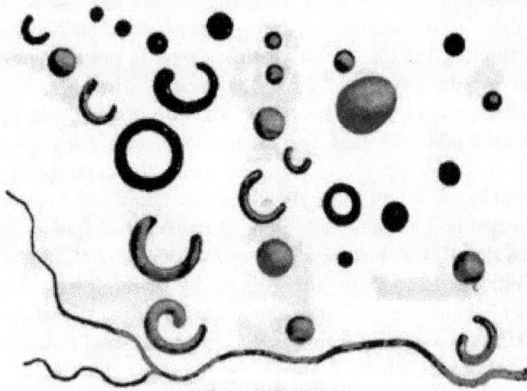

Fig. 2.—Burial Mounds.

works have been found in the Ohio Valley, in the more northern States,
west of the Mississippi and in Michigan. Upon excavating the more
finished mounds of the group described, they were found filled with
ashes, mingled with charcoal: some of them to the depth of twenty
inches below the surrounding surface of the soil. In this group were
found two mounds much larger than the others, (one is represented
in the engraving), shaped like the body of a tortoise, known as battle
mounds, and said to contain the ashes of hundreds slain in battle. Both
these mounds were found to be filled with ashes and charcoal like the
others, thus confirming their traditional history.

About two hundred and fifty yards south of these mounds, another
group of finished works was found, where the bodies were deposited in
the more usual manner without burning.

2

These two modes of burial, so widely different and in the same locality, mark either a sudden change of custom or the presence of two distinct races at different periods of time. Tradition asserts that there was such a sudden change of mode of burial in obedience to the command of the prophets, for the reason that, while the people were burning the body of a great and good king, suddenly the sun (their chief deity) refused to shine, although there was not a cloud in the sky. This was taken as a sign of disapprobation of the custom, which gradually ceased thereafter.

It has been generally supposed that those mounds, which showed the frequent or long-continued action of fire, were used for sacrificial purposes only. It seems however more likely that these cinerulent structures were simply the depositories of the bodies of the dead, and this the traditions affirm.

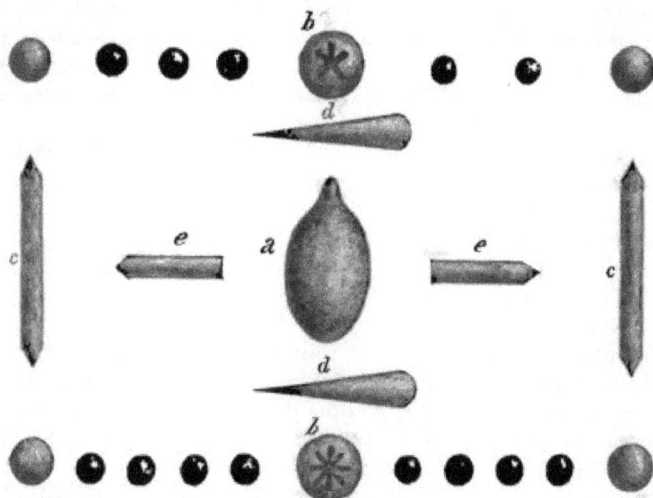

Fig. 3.—A Royal Cemetery.

The second group noticed in this connection is more complicated (Fig. 3), and presents a greater variety of forms. It is found (or was in 1840) on the north side of St. Peter's river, about sixty miles above its junction with the Mississippi, in what was then the Territory of Minnesota.

It is thus described: The central embankment, in the form of the body of a tortoise, is forty feet in length, twenty-seven in breadth, and twelve in perpendicular height. It is composed in part of yellow clay, brought from some distant place. The two pointed mounds north and south of this are formed of pure red earth, covered with alluvial soil. Each is twenty-seven feet in length and six in height at the largest end, gradually narrowing and sinking at the top until they terminate in a point. The four corner mounds were each twelve feet high and twenty-five in diameter at the base. The two long mounds on the east and west sides of the group were sixty feet in length, twelve feet in diameter at the base, and eight feet in height. The two mounds on the immediate right and left of the central effigy, were twelve feet long, four feet high, and six in breadth. These were composed of sand, mixed with small bits of mica to the depth of two feet, covered with white clay, with a thin layer of surface soil on the top. The large mound in the center, south of the effigy, was twelve feet high, twenty-seven in diameter, and composed of a stratum of sand two feet in depth, covered with a mixture of sandy soil and blue clay. The similar work on the north of the tortoise was of like formation, four feet high and twenty-two feet in diameter. Thirteen small mounds whose dimensions are not given, complete the group.

Only a glance at this cluster of mounds, twenty-six in number, presenting such variety of forms and peculiar arrangement, and which must have required so much time and labor for their construction, is needed to convince the observer that they were intended to perpetuate some history, and that each of the hieroglyphic symbols of which the group is composed had its special significance, which was well understood by the builders and their cotemporaries.

But what was that history, or what event is recorded here? The works themselves give no answer. Tradition asserts, that this was the royal cemetery of a ruler known as the Black Tortoise, and was designed to commemorate the title and dignity of a great king or potentate. The tortoise-shaped central mound (a) was his tomb. The four corner mounds were called Mourning Mounds. The two larger mounds (bb) directly north and south of the effigy were the burial places of chiefs. The number interred in each is recorded in the number of small mounds on each side of them—five in the northern and eight in the southern line. The two long embankments (cc) at the extreme right and left of the works, were known as points of honor, and are said never to occur except in connection with those works which symbolize royalty. The two pointed mounds (dd), and described as twenty-seven feet long, six feet in width at the larger end, tapering down from the top and sides to a vanishing point, are known as mounds of extinction, and tell us that he

was the last of his line.   These too are never found alone, but always in
connection with larger works.   The mounds (*ee*) on either side of the
central effigy are the burial places of prophets.   In these it will be
remembered small bits of mica were found mingled with the ashes.   The
presence of this substance in a certain class of mounds, in localities so
remote from each other, from Minnesota to the Scioto Valley—sometimes in large circular plates, but oftener in countless smaller fragments,
has called forth much speculation as to its use by the ancient inhabitants.
It has been suggested that it may have been used for mirrors, or again
for ornament, or, on account of its preciousness, as a medium of commercial transactions.   But when it is remembered that it is never found
indiscriminately with other deposits in many mounds of the same group,
we may safely conclude that it was set apart for a special use.   Tradition
says that it was sacred to the prophets, and was deposited in their tombs
alone ;—that they had the mysterious power of calling fire from heaven,
which was distributed to the minor prophets by whom the sacred fires
were kept perpetually burning ; that the fire used at the annual feast
in their most holy places was thus received from the sun upon the
summit of the sacred altars.   This bringing fire from heaven is found in
classic stories and in the traditions of many lands, as every school-boy

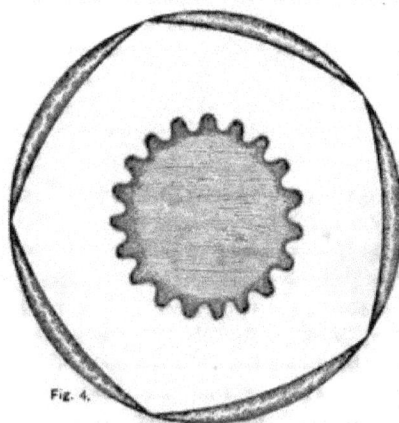

Fig. 4.

knows.   So Zoroaster taught his
disciples, that the sacred fire
which he committed to their
care had been brought direct
from heaven.   " It is possible
that the prophets of the ancient
Americans were able in some
manner to construct lenses from
plates of mica, of sufficient power to ignite the fuel upon the
sacrificial altars." [1]   The Mexi-
cans in ancient times called obsidian "the shining god," and
held it in high estimation.

Several works have been ob-
served of the form shown in Figure 4.   The one here represented
is described as located on the lowlands of the Kickapoo River in
Wisconsin.   The central work, with radiating points, sixty feet in

[1] Pidgeon.

diameter and three feet in height. This is inclosed by five crescent-shaped works, having an elevation of two feet, and all presenting a level surface at the top. It is traditionally represented to have been occupied only during sacrificial festivities consequent upon the offering of human sacrifices to the sun, which the central mound was said to represent. Upon excavating, after removing the soil from the top, the central portion, for a space twelve feet in diameter, is found thickly studded with plates of mica set in white sand and blue clay; and, says the observer "had this surface soil been removed with care, and the stratum beneath washed by a few heavy showers of rain, under the sun's rays it would have presented no unapt symbolical representation of that luminary." The sacred Pentagon, Fig. 5, is found in close proximity.[1]

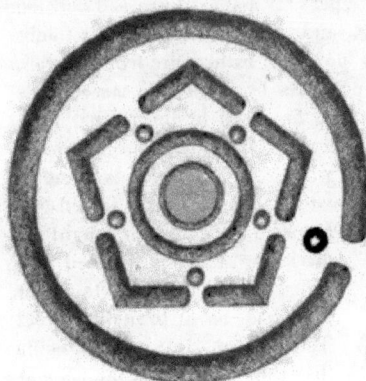

Fig. 5.—Sacred Pentagon.

As before stated, no class of works has awaked more curiosity, or elicited more unsatisfactory speculations, than these animal effigies; and among these the most singular and enigmatical are those representing the larger animals, and the human form on a gigantic scale, and generally with such accuracy of delineation as to leave no doubt as to what particular animal was intended to be represented by the figure. Sometimes these huge representations of beasts, birds and men are grouped together in such strange and grotesque combinations as to forbid all attempts to discover the design of the builders in their erection. A few of the most common forms are shown in the accompanying engravings.

That the mastodon is intended by figure 1 is conceded by all—as far as known—who have described it. I am not aware that it has ever been found outside of Wisconsin. There it frequently occurs, either

[1] This is represented here because of its intimate relation to the one just described, which is found associated with it. The outer circle is twelve hundred feet in diameter, in the center is the sacrificial altar, upon which human sacrifices were said to have been offered twice a year. In the spring the oldest man of the nation, willingly—so great was the honor—presented himself as the victim. In the autumn a female was sacrificed. If the day was cloudy, the offering was left upon the altar of sacrifice until the sun looked down upon it, which was considered a sign that the sacrifice was accepted. The people then repaired to the festival circle with rejoicing, where the feast was celebrated

alone or in companionship with other mounds. As men in all ages, in
their first attempts at pictorial art, have been accustomed to delineate
only those objects which were most striking and with which they were
most familiar, we may well believe that the ancient Americans were not
unacquainted with this king of beasts, and that they lived in those days
when those gigantic animals roamed over the plains in vast numbers,
whose skeletons have been so often found in Missouri.

Fig. 6.
Bird and Beast.

The combined figures of bird and beast as represented at Fig. 6 are also of frequent occurrence, parti-
culary in Wisconsin. The one here delineated is one
hundred and eighty feet in length, and forty-four in its
greatest breadth. The whole is composed of reddish clay,
but covered to the depth of twelve inches with a black
alluvium. It was designed to record the change in
title of a sovereign line of rulers. The head of the
beast being merged in the body of the bird concedes
to the conqueror the right of dominion. The two
truncated mounds, one on each side of the beast, record
the extent of his humiliation. They are altar mounds, on
which were sacrificed his descendants both male and
female.

The effigy shown in Fig. 7 is unmistakably human. It memorializes a
hereditary chief of royal line, but who, according to the record, could
not yet have been a sovereign ruler, as no mound of honor indicating that
condition is found in connection with it. He was thus memorialized
because he fell in battle, and with him his son, whose memory is perpetuated in the truncated mound between his feet.

The amalgamation group (Fig. 8) is more complicated and enigmatical, and but for the traditions concerning it would doubtless always
remain so. The beast is one hundred and eighty feet in length; the
human effigy perpendicular to it is one hundred and sixty. On either
side of the horizontal figure is a truncated work eighteen feet in diameter
and six feet in height. The summits of both are flat. The representations
of horns, which are very distinct, are of different dimensions. The
main stem of the front horn is eighteen feet in length. The one
which inclines backward is twelve, the longest antlers are six, and
the shortest three feet in length. At the foot of the human effigy
is attached an embankment running parallel with the horizontal figure,
eighty feet in length, twenty-seven in diameter and six in height. On a
line with this is a series of conical mounds, the largest of which is also

twenty-seven feet in diameter and six in height. From this the others diminish on either side and terminate in mounds eighteen feet in diameter and three in height. The group thus described is represented to have been erected to commemorate an important event in the history of two friendly nations, which were once great and powerful, but now reduced by long-continued wars against a common foe; and being now no longer able to maintain a separate national existence, they resolved to unite their forces under one title and sovereign. One was known as the Elk nation, the other was the Buffalo. This work was designed as a public record and seal of their amalgamation. This fact is plainly expressed by the union of the head of the Buffalo with that of the human effigy representing the sovereign of the Elk nation, and also by the joining of the hand of the one with the foot of the other.

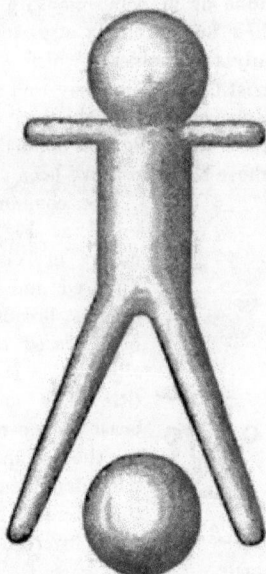

Fig. 7.—Human-Shaped Mound.

Horns appended to effigies represent warriors; their length and number the relative power of the two nations at the time of their union. The Buffalo was therefore manifestly recorded as the weaker of the two, as his antlers are seen to be smaller and in a declining position. The fact is also here recorded that, when the union was fully consummated the nationality of the Buffalo became extinct. This is shown by the presence of the mound of extinction

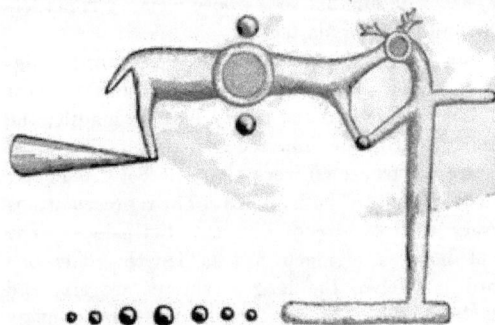

Fig. 8.—Amalgamation Group.

—before described—in connection with the Buffalo and terminating at his hind feet. The two truncated mounds on either side of the

animal effigy are sacrificial altars upon which appropriate sacrifices
were offered, not only at the time of the erection of the works,
but annually thereafter ; the fires of which were kept burning
until the smoke from both united in one column above the
mound. This annual sacrifice symbolized the renewal of the covenant
entered into when the compact was made. The seven truncated mounds
in a line with the embankment upon which the human figure stands,
(and known as a symbol of nationality) are matrimonial memorials, and
record the international marriages of seven chiefs which occurred during
the construction of the work, and which were also a further ratification
of the national union here perpetuated. Upon excavating the altars,
after the alluvial soil was removed, a stratum of burned earth mingled
with ashes and charcoal was disclosed, to the depth of fourteen inches.
This group was found upon the northern high land of the Wisconsin
River, about fifty miles from its junction with the Mississippi.

In that part of the work where the heads of the two effigies unite, an
oak was standing at the time of its first examination. Upon a second
visit it was not there, but the stump showed by its concentric annual
rings of growth that it was four hundred and twenty-four years old.
Works of this description, which occur so frequently in Wisconsin, have
also been observed in Northern Illinois.

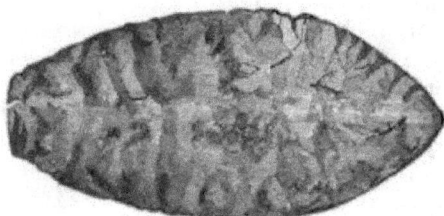

Lance Head..

CHAPTER IV.

The preceding remarks upon the general subject of Archæology, with the few notices of traditions concerning the ancient inhabitants of America, are all that the limits of this article will permit, as well as all which our present purpose demands. Nor has it seemed necessary to describe those extensive and imposing works, which are found scattered through the Central States, from the lakes to the Gulf of Mexico, and especially in the Ohio Valley, consisting of walled towns, embankments enclosing large areas of land, in squares, circles, octagons and the like, associated with mounds of prodigious size; for these have been so often described and delineated that whatever comparison of them with the monuments of Missouri may be thought desirable may be readily accomplished by reference to the works of those authors, who have published so many valuable descriptions of these antiquities, and which are to be found in almost every public library.

That Missouri was once the home of a vast population composed of tribes who had fixed habitations, dwelt in large towns, practiced agriculture on an extended scale, with a good degree of method and skill; who had also a well-organized system of religious rites and worship, and whose æsthetic tastes were far in advance of the savage tribes who roamed over her prairies and hill ranges when her great rivers were first navigated by the white men, is, I am confident, no difficult matter to prove. Says Mr. H. M. Brackenridge, who was an extensive traveler, and a man of excellent judgment, in speaking of the ancient works in the Mississippi Valley: "It is worthy of observation, that all these vestiges invariably occupy the most eligible situations for towns or settlements; and on the Ohio and Mississippi they are most numerous and considerable. There is not a rising town, or a farm of an eligible situation, in whose vicinity some of them may not be found. I have heard a surveyor of the public lands observe, that wherever any of these remains were met with, he was sure to find an extensive body of fertile land."

Although, for more than three-quarters of a century since that time,

the waves of an advancing civilization and the hand of agriculture have
passed over them and utterly destroyed vast numbers, including many
of the most remarkable ones, which arrested the attention of every
beholder,—still, any one at all familiar with those which now remain
would write the same things to-day. The name of the city of St. Louis
was once Mound City, called so on account of the number and size of
those ancient works which once stood upon her present site. The larger
of them are all demolished, while the few which yet remain are so small
that they would hardly be noticed save by the eye of a practical observer.
The same may be said of nearly all which once crowned the terraces of
the Mississippi along her eastern border, and those of the Missouri and
her tributaries.

Notwithstanding all this widespread demolition and obliteration, there
is doubtless now no richer field for archæological research in the great
basin of the Mississippi than is to be found in the State of Missouri.
As has been already stated, the most important works are found
located in the vicinity of extensive areas of fertile lands, and upon the
most eligible sites for towns and cities. The same locations would
naturally be the first to be occupied by the pioneer settlements of our
own times, and these aboriginal remains would be the first to be obliterated. It is not surprising, therefore, that the earlier notices of the
ancient monuments of this valley are so meagre and unsatisfactory,
especially when we remember the peculiar vicissitudes of a frontier life,
which necessitated unceasing toil and eternal vigilance : continually men-
aced, as the early settlements generally were, by a wily, savage foe.

It should also be remembered that until quite recently the prevailing
opinion concerning mounds and embankments was that they were the
work of the red men, and to this day they are known among the masses
as Indian mounds.

Notwithstanding the fact that multitudes have been destroyed, there
still remain so many vestiges of an ancient race—not only upon the
alluvial plains of our larger rivers, but also in the interior valleys,
watered by smaller streams and rivulets, and also upon the sterile slopes
and summits even of the Ozarks—that Missouri still presents a most
inviting field for the labors of the archæologist. A proper examination and description of them all would involve no inconsiderable
expenditure of time and money, and require a volume for their
elucidation. It cannot therefore be expected that we can do more
in this article than to describe the different classes of those remains—
with their most prominent characteristics—which are best known and

which have been the most thoroughly explored. In carrying out this design, it will perhaps best serve our purpose in the way of method and convenience to consider them under the following general divisions : 1st, Sites of towns or cities. 2d, Burial mounds, caves and artificial caverns. 3d, Sacrificial or temple mounds. 4th, Garden mounds. 5th, Miscellaneous works. 6th, Pottery ; and 7th, Crania.

I.—Sites of Towns or Cities.—The early French explorers of the Mississippi and Missouri rivers, and the territories through which they flow, seem to have taken no notice of the ancient monuments along their course ; or if they did, they doubtless ascribed their origin to the red men, who were found occupying, in some instances works of similar construction.

But when permanent settlements had been established along their banks, with the consequent increase of travel, these works ere long attracted the attention of the historian, and awakened an interest which resulted in their more careful examination. The early writers, as they became familiar with the habits and social condition of the Indians, and in view of the magnitude of the structures they so frequently met with, as well as the skill and herculean labors required for their erection, make frequent mention of their doubts as to the ability of the Indians to erect monuments of such prodigious proportions. And not until St. Louis became an incorporated town, and the capital of that vast extent of territory then known as Upper Louisiana, do we find any descriptive accounts of the ancient works which at that time occupied the terraces upon which this great city now stands.

Notwithstanding the meager and unsatisfactory character of the accounts which have been preserved, still, we are thankful for the crumbs of information the early observers have left us, and will endeavor to make the most of them.

Mr. H. M. Brackenridge,[1] writing in the year 1811, says: "I have

---

[1] The work of this author (" Views of Louisiana ") seems to have been the perennial fountain from whence many subsequent writers upon American Archæology, both in this country and in Europe, have drawn much of their inspiration and many of the facts and germinal suggestions which they have elaborated with extended speculations, and frequently without any mention of their obligation to this writer for the facts and suggestions which have been so freely made use of. Mr. Brackenridge, I believe, was the first American author who alludes to the statements of Plato concerning a people who had come from an island in the Atlantic, in great numbers, and overran Europe and Asia, and known as the Atlantides, which island was said to have been sunk by an earthquake 9000 years before his time. He notes, also, a similar tradition among the Romans, and thinks it possible America may have been referred to.

frequently examined the mounds at St. Louis. They are situated on the
second bank, just above the town, and disposed in a singular manner;
there are nine in all, and form the three sides of a parallelogram, the
open side towards the country being protected, however, by three
smaller mounds, placed in a circular manner. The space inclosed is
about three hundred yards in length and two hundred in breadth.
About six hundred yards above these is a single mound, with a broad
stage on the river side; it is thirty feet in height, and one hundred and
fifty in length; the top is a mere ridge of five or six feet wide. Below the
first mounds there is a curious work called the Falling Garden. Advantage is taken of the second bank, nearly fifty feet in height at this place,
and three regular stages or steps are formed by earth brought from a
distance. This work is much admired—it suggests the idea of a place
of assembly for the purpose of counselling on public occasions."

Accompanying the foregoing description is a simple diagram which,
as it does not seem to be the result of any actual survey, and therefore
of no scientific value, need not be reproduced in this connection.

Dr. Beck, who noticed them twelve years afterwards, presents in his
work another diagram, which seems to have been the result of more
careful observation, although in this, however, one of the nine, and the
three smaller mounds described by Mr. Brackenridge as protecting the
side of the parallelogram opening towards the country, are wanting.
From all the information I can gather, I believe the following plan will
present the true relation of the mounds here described:

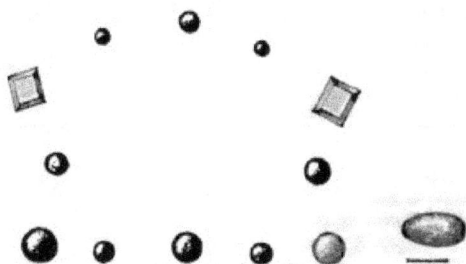

Diagram of St. Louis Mounds.

THE BIG MOUND AT ST. LOUIS.

One of the above group undoubtedly represents the old landmark known as the Big Mound, (a representation of which as it appeared at the time of its removal, faces the first page of the present volume), which once stood at the corner of Mound street and Broadway, but which was entirely demolished in 1869. This I suppose to have been the terraced mound, represented by Mr. Brackenridge to have been located six hundred yards north of the main group. The Big Mound is known to have been beautifully terraced, and nothing of the kind is mentioned in connection with those constituting the parallelogram. Nor is the Falling Garden spoken of as a mound, but only as a terraced bank. For these and other reasons which need not be dwelt upon, after much reflection, I am persuaded that the terraced mound, afterwards known as the Big Mound, was the last to disappear before the encroachments of the rapidly-growing city. Be this as it may, this most interesting work will be particularly described under the more appropriate head of Sepulchral Caverns, when I shall be able to speak with more confidence, as I shall give there the result of my own observations. There were formerly many other mounds in the immediate vicinity of St. Louis, rivaling in magnitude and interest those described by the authors just quoted, but which escaped their notice In fact, the second terrace of the Mississippi, upon almost every available commanding point of elevation, was finished with them. Nineteen years ago, in a conversation with the late Col. John O'Fallon, he informed me that his family residence on the Bellefontaine road was erected upon one of those ancient mounds. It must have been very large, although I do not recall the dimensions. He stated, further, that as the summit was being leveled, preparatory to building, human bones by the cart-load were disclosed, along with stone axes and arrowheads and the like, without number. He then led me to the forest west of his dwelling, and called my attention to the small hillocks which abounded there in prodigious numbers, which he conjectured were the residence sites of former inhabitants, because of their regularity, and from the fact that upon excavating them they disclosed ashes and charcoal.

Still farther north, upon the highest points of the second terrace, I have traced the remains of others which must have been quite imposing before they were subjected to the leveling influence of agriculture. In Forest Park, a few miles west of the city, there is a small group of mounds which the park commissioners, I am happy to know, have resolved to preserve. It is a pity that none of the larger ones have been spared, to stand hereafter as the memorials of a people whose origin is

hid in the night of oblivion. But let them remain, such as they are, and
when future generations shall throng the green groves and shady walks
of that beautiful garden of their great city, these shall recall the fainting
echoes of another race, whose homes once clustered, in days long gone,
upon the banks of that great river where a statelier—can we say happier
—city stands to-day.

The works thus briefly noticed are only a few of the great group of
large circumference, of which that king of mounds, on the fertile plains
across the river, known as Monk's Mound, was the radiating center.
That high place was a temple mound—the holy mountain for this whole
region, doubtless,—and the smoke which ascended from the perpetual
fire of its sacred altar could be seen for many miles on every side.

But while our business now is with the ancient people of Missouri, it
should be borne in mind that the imaginary lines which divide us into
States had no existence in those other times, when a mighty people
dwelt upon either side of the Mississippi, outnumbering far, perhaps,
the present occupants; who were homogeneous in their commercial
pursuits, arts and worship. They traded with the nations who dwelt by
the sea, and brought from thence the shells and pearls of the ocean, and
left them in their tombs, along with the precious wares of their own
handicraft, for our admiration and instruction.

But before we leave St. Louis, another work demands a notice,
which the following (Fig. 9), will illustrate.

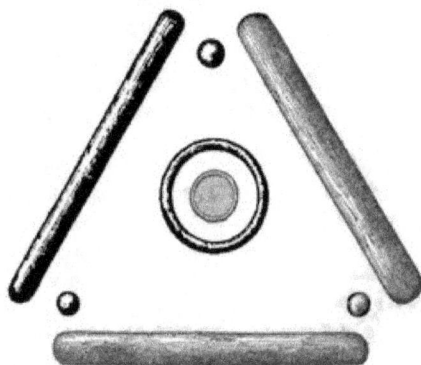

Fig. 9.—Historical Mound.

This class of works
appears frequently in Iowa,
but was formerly found in
greatest numbers in Mis-
souri. The one figured
here was located on Root
River, about twenty miles
west of the Mississippi.
The central mound is represented as being thirty-six
feet in diameter, and
twelve feet in height. The
circle inclosing it was
nearly obliterated. The
long embankments which
form the sides of the triangle were each one hundred and forty-
four feet in length, and respectively three, four and five feet in

height, and twelve feet in diameter; and what is singular, the sum of the heights of the embankments equals the vertical height of the central mound, and these two amounts multiplied together, give the exact length of the embankments. Sometimes works of this description are built in the form of a square, with four embankments; but of whatever form, it is stated that the same relation of the sum of the heights of all the embankments to the height of the central mound is always presented, and the product of these gives the length of the embankments.

A group precisely similar to the one just described, and of large dimensions, once stood near the village of St. Louis. Its precise location cannot be learned, as it was demolished somewhere between the years 1835–40. This class of mounds will be further noticed under the head of Miscellaneous Works.

The evidences of a dense pre-historic population in Missouri are nowhere so abundant as in the southeastern counties of the State. These consist of mounds of various dimensions and forms, sometimes isolated, but oftener in groups of peculiar arrangement; also embankments and walls of earth inclosing large and small areas, in which may be traced the lines of streets—if such they may be called—of a village or city, and numberless sites of former residences. One of the largest mounds in this region, is about four miles from New Madrid, and, as described in 1811, is twelve hundred feet in circumference, forty feet high and surrounded by a ditch, five feet deep and ten feet in width. New Madrid was unquestionably once the great metropolis of a vast population, the remains of whose villages are everywhere met with, upon the banks of the numerous bayous which abound in the several counties in this portion of the State. For the reason before mentioned, one group only can be particularly described.

The one selected is situated upon Bayou St. John, about eighteen miles from the town of New Madrid. The bayou at this point is one mile and a half in width; its whole length may be stated in round numbers to be about seventy-five miles. While, in the notices of the earlier travelers, it is described as a lake with a clear, sandy bottom, it is now a sluggish swamp, filled to a great extent with cypress trees.

Upon the western bank of the bayou the works to be described are located. They consist of inclosures, large and small conical and truncated mounds in great numbers, and countless residence sites of the ancient inhabitants. From the level of the bayou to the prairie land above, the ascent is by a gradual slope to a vertical height of fifteen feet. Upon this belt of sloping ground, now covered with a heavy growth of timber, the works are most numerous ; while from its edge, westward, the level prairie (that is, the alluvial plain of the Mississippi) has been under cultivation for sixty or seventy years. Here, including forty acres of the cultivated field and ten of the sloping timber belt, is an area of about fifty acres, enclosed by earthen walls which may be distinctly traced for several hundred feet, but gradually disappear on the western side, having been nearly obliterated by the long cultivation of the field. Where it is best preserved in the timbered land, its height was found to be from three to five feet, and fifteen feet wide at the base. [1] In the centre of the western side of the enclosure and close to the wall, is a mound of oblong shape, three hundred feet in length at the base, and at its northern end one hundred feet wide, and twenty feet high at the present time, as near as could be estimated by careful stepping. The top of it slopes gradually to the south, and although the plow has passed up and down its sides for sixty years, still on its eastern side may be distinctly seen the evidences of a graded way to its summit. Close to its northeastern side, where the mound is widest, is a deep depression in the field, about ten feet in diameter. Mr. Wm. M. Murphy, a farmer who has long resided in the neighborhood, told me that when he first saw it he could not get in and out of it without a ladder, and that it had since been nearly filled up by the tillers of the soil with stumps, logs and earth. In the centre of the enclosure stands a circular mound seventy-five feet in diameter, and also twenty feet high, which upon examination disclosed nothing but broken pottery. It belongs to that class usually termed residence mounds. The view from its summit towards the west and south commands a prospect several miles in extent ; on the north the view is cut off by a heavy growth of timber, and on the east by the cypress swamp. In a direct line with the two mounds thus described, partly upon the edge of the cultivated field and partly upon the declivity which descends toward the swamp, in the midst of a group of smaller works, stands a large burial mound, twelve to fifteen feet in height, and one hundred feet in diameter. Its original height could only be conjectured, as it has long been occupied as a residence site by the present inhabitants. The ruins of a log house are still standing upon its summit. It has been the sepulchre of many hundreds, perhaps a

---

[1] It will readily be perceived that absolute accuracy of measurement would be impossible, where the ground has been so much disturbed by cultivation.

3

Small vessels of Pottery, Stone Pipe, Stone Implements and discoidal Stone from New Madrid, Mo.

thousand individuals. The manner of interment, as far as my own observations extended, was to place the corpse upon the back, with the head towards the centre of the mound; the vacant space between each deposit being generally two or three feet. When the inner circle was full, another would be formed outside of it. In two burial mounds in this region, which were only from three to five feet in height, and fifty or sixty feet in diameter, I found this process of burial continued far beyond the circumference of the mound; in which cases the graves had been dug in the natural bed of the plain upon which the mound was erected, and were generally from three to four feet in depth. The kind of pottery found in these is precisely similar to that taken from the centre of the mound, and was always in the same relative position to the skeleton. Three vessels were usually found with each individual. Two were water jugs, and placed on each side of the head; the other, a receptacle for food, rested upon the side of the chest, and was kept in place by the angle of the arms, which were folded across the breast. These vessels will be more particularly described hereafter.

Within the enclosure before described, beginning near the margin of the bayou, extending up the side of the declivity, around the burial mound, and continuing quite a distance into the inclosure, are great numbers of depressions, or shallow pits in the soil, from one to three feet in depth and from fifteen to thirty in diameter; sometimes in parallel rows, and usually about thirty feet from centre to centre. In many of these, forest trees of large size are still growing, and others equally large are lying upon the ground in various stages of decay. Upon digging into them, almost every shovelful of earth disclosed pieces of broken pottery; many of these fragments indicated vessels of large size which must have had a capacity of from ten to fifteen gallons. Upon joining the fragments together, the mouths or openings were found to vary from three to twelve inches in diameter. They were doubtless stationary receptacles of food or water, as they were so thin that it would hardly seem possible they could be moved, when filled, without breaking. In many of these depressions were observed large rough masses of burnt clay, of the color of common brick, full of irregular and transverse holes, which seem to indicate, that, before it was burned, the desired form of a chimney, or oven, had been rudely made out, by intertwining sticks, twigs and grass, and the whole plastered inside and out with moist clay, to the thickness of several inches, and then burned until it became red and nearly as hard as the bricks now in use. At the depth of about two feet, at the bottom of all which were examined, what

seemed to have been a fire-place was disclosed. The earth was also
burned, so as to present the color and hardness of the fragments of brick,
to the depth of several inches. Along with the broken pottery were
found, quite often, fragments of sandstone of various sizes, the larger
pieces with concave surfaces, and all showing that they had been used for
polishing or sharpening purposes, especially the smaller ones, which are
covered with small grooves one-eighth of an inch deep across the whole
length and width, and at various angles with each other, as though they
had long been used for sharpening some small metallic instrument or
graver's tool.

Water Jugs and Food Vessel.

Another interesting and suggestive feature of these works is worthy of
notice. Along the shore of the bayou, in front of the enclosure, small
tongues of land have been carried out into the water, from fifteen to
thirty feet in length by ten to fifteen in width, with open spaces between,
which, small as they are, forcibly remind one of the wharves of a sea-
port town. The cypress trees grow very thickly in all the little bays
thus formed, and the irregular, yet methodical, outline of the forest,
winding in and out, close to the shore of these tongues of land, is so
marked as to remove all doubt as to their artificial origin. Although the
channel of the Mississippi is now from fifteen to eighteen miles east of
this point, there is no question that this long bayou was one of its ancient
beds. It is well known that at New Madrid the river has receded at the
rate of one mile in seventy years. With the supposition that its recession
has been uniform, at this rate nearly a thousand years must have passed
since the Mississippi deserted the banks upon which these works are
located. But this, could it be proven, would give us no positive testimony concerning their age. When the river changed its course, a lake

took its place. The change therefore must have been somewhat sudden, for according to its prevailing habits, while it wears away the shore upon one side it leaves a corresponding deposit of alluvium upon the other.

The numerous miniature wharves would suggest that the inhabitants were fishermen and had plenty of boats of some sort, which being so, these waters must have been navigable and not filled up as now with an almost impenetrable cypress forest.

Large Water Vessel.

While it is true that the most important works are all situated upon the high ground, fifteen feet above the water level, some of the smaller ones are located upon the intermediate declivity, and near the shore of the bayou, as also some of the residence sites.

If we assume their occupancy to have been contemporaneous with the presence of the river, they would be subject to overflow by the annual floods, and the wharves would be swept away. It seems probable therefore that the time when they were occupied was long subsequent to the change in the course of the river. The idea of the great antiquity of these works, entertained when I made the report of their examination, to the St. Louis Academy of Science, I confess has since been somewhat shaken, the reasons for which may appear as we proceed. I am reminded however that, for the work of which these are the initial chapters, a picturesque and, so to speak, a topographical description of the ancient monuments of Missouri is desired, rather than a dry detail of facts with extended generalizations. Considerations therefore which might otherwise be appropriate in this connection will be reserved for a more fitting opportunity.

One mile south of the remains under consideration, and about three hundred feet from the margin of the bayou, is a peculiar work, in the form of an oval or egg-shaped excavation, one hundred and fifty feet long in its largest diameter and seventy-five feet wide and about six feet deep. It is surrounded by an embankment about eight feet in height around its northern curve: on the southern end the wall is not over five

Small Drinking Vessel, and Stopper.

feet, in which is a narrow opening, and extending from it is a curved, elevated way to the swamp, in which the earth taken from the excavation seems to have been deposited, until a circular mound or wharf was raised about twenty feet in diameter and five feet high in the centre. The same opening and elevated way is seen at the northern end, extending to the water. It is doubtless an unfinished work, but its purpose cannot be conjectured.

About eight miles, in a southeasterly direction, from the works upon Bayou St. John, upon what is known as West Lake, is another extensive group almost identical with those described above, differing chiefly in this, that they are covered throughout with a heavy growth of timber; and the residence sites are found covering a much larger space, and in prodigious numbers; while in the center of the group is an open space of several acres which seems to have been made perfectly level, containing no elevations or depressions whatever save what may have been produced by the uprooting of timber.

The aboriginal remains thus briefly described are only small groups of the multitudinous works with which this whole region abounds, and in many instances are still covered with the primeval forests.

They seem to increase in number and size as we approach the town of New Madrid, where they appear in structures of much greater magnitude, one of which has been already noticed. Their character at this place would seem to indicate that here was the seat of government and commercial metropolis of a dense population, which occupied a large extent of territory, embracing not only New Madrid county, but also the counties of Mississippi, Scott, Perry, Butler, Pemiscot, Scotland, Madison, Bollinger and Cape Girardeau, all of which contain the same class of works, and whose authors were the same people. Further explorations, I have no doubt, will disclose their presence in other counties adjoining.

CHAPTER V.

Notwithstanding the variety of form presented in the multitudinous structures throughout the continent of North America, the comparison of many of their most prominent characteristics makes it reasonably certain that one people were the authors of them all. While many of them in the order of their age belong to periods more or less remote, reaching back many hundreds and perhaps thousands of years, many others are comparatively recent. Taken as a whole, the thoughtful observer will see, in this diversity of configuration and grouping, that natural order of growth which might be looked for in the slow development of a national life, whether generated among the people themselves or helped forward by occasional and accidental impulses from without. It seems highly probable that there were two slowly-moving streams of migration from the north; the most important one on the east of the Mississippi, the other through the territories lying west of the river. This southward movement of a vast people seems to have been arrested in the valley of the Ohio for a long period of time. Otherwise the fact can hardly be accounted for that here occur the most stupendous monuments of their industry and skill, and also the most striking evidences of the stability and repose of their national life. Here the mound-builders reached the highest stage of civilization they ever attained this side of Central America and Mexico. The movement upon the western side of the river, while it had its source in the one great fountain-head at the north, does not seem to have been so well defined in all its characteristics, notwithstanding the fact that the population in Missouri at one time was as great, and, we have reason to think, greater than in Ohio. The cause may have been that they never enjoyed a season of repose and exemption from war to such a degree as to render it possible for them to devote the time and concentrate their energies upon their internal affairs to the extent which resulted in the more advanced civilization of the eastern tribes. There seems to have been one prevailing system of religion among them all, which was based upon the worship of the heavenly bodies. This remark applies not only to people of North America, but to the ancient inhabitants of the southern continent as well. The temple mounds in both, though built of different materials, are the same in form and purpose.

While the oneness of their forms of worship of itself proves nothing as to the unity of their origin, still, when taken in connection with the fact of their constant intercourse, and the identity of so many rites and customs among them all, it is believed no extended argument is needed, as before stated, to prove that, whatever may be the relative age of the groups of works found in different localities, they were all built by one people. In view of the foregoing it ought not to be surprising if, as we trace the history of their development as recorded in their remains, we find here and there traces of a radical change in some of their customs. The one we have now to consider is a most important and significant one, which relates to the disposition of their dead. This has already been noticed (see p. 17, fig. 2), as illustrated in the two cemeteries in Carroll County, Illinois, with traditional reasons for the substitution of mound burial for cremation. Many able writers upon American antiquities have given much attention to the numerous class of works which have usually been denominated sacrificial mounds.

These are described as presenting upon excavation a basin-shaped cavity of varying dimensions: frequently paved with stones, and containing ashes and charcoal, which are sometimes mingled with various implements and ornaments, all showing the action of fire. To my own mind the evidences are almost conclusive that these should be denominated Cremation Mounds; and that up to a certain period this was the usual, and perhaps, universal, method of disposing of the remains or departed friends. The size of the mound would then indicate the rank of him whose body was thus consumed therein. Upon no other hypothesis can we account for the earth being heaped upon the so-called altars while the fires were yet burning, leaving some portions of the wood yet unconsumed. At length this practice ceased and mound burial took its place. The latter custom seems to have been the one universally practiced by the mound-builders of Missouri.

While cremation mounds occur in Iowa and Wisconsin, if any exist in Missouri they are yet to be discovered. But here even the mode of burial was not uniform throughout the State, nor always in the same locality even. One class, in the bayou St. John group, has already been described. It is to be remembered that in these no implements whatever were found with the interments—nothing save the earthen vessels for food and drink. Occasionally a flint spear and arrowhead would be disclosed, but in such relations that I have no doubt their presence was accidental. These mounds I believe to have been the ordinary burial places of the people. In others, as was the case with the one upon which the O'Fallon mansion stands, great numbers of stone axes, arrow-points, and the like abound.

In the one case, only those domestic utensils were deposited which minister to the comfort of their domestic life; in the other, those which served them in war and manly activities. Nor does this seem strange, when we remember the belief, so common among mankind in certain stages of civilization, that those pursuits to which the individual was devoted in this life are continued in the life beyond the grave; consequently, if he had been a great hunter or mighty in war, it would be most natural to deposit with him, in the tomb, his arms. But if the nation were at peace, and unused to the arts of war, his friends would think only of a necessary supply of food and drink; hence vessels of pottery would be the sole accompaniments of his journey.

Should the idea here advanced be substantiated by future investigation, that cremation was once the prevailing custom and that at some period it was discontinued and mound-burial adopted in its place, then it would seem altogether probable that Southeast Missouri was peopled at some time subsequent to that event, and therefore the works so abundant there are more recent than those of the Ohio Valley.

Another class of sepulchral mounds, whose occurrence is somewhat rare, has been observed more particularly in the Western Central States. Generally they are of large dimensions and contain a chamber or vault, which is sometimes rudely finished with stone. The floor is usually on a level with the natural surface of the soil, upon which the dead were placed, in a reclining posture. The most conspicuous example of this class is the one known as the Big Mound, which once stood at the corner of Mound street and Broadway in St. Louis, but which, as before stated, was removed in 1869. A representation of it, as it appeared, is given in our frontispiece.

Of all sepulchral mounds thus far examined, this was the king. If its magnitude, or rather the size of the vault within it has any significance, it would seem to have been the tomb of the most holy prophets or of the royal race. The statements concerning its dimensions are widely different. According to one observer, it was four hundred feet in length, two hundred feet wide at the base and over fifty feet high. According to Mr. Brackenridge, it was one hundred and fifty feet in length and thirty in height. The latter figures are probably not far from the truth.

These discrepancies are not difficult of explanation when it is remembered that in its construction, advantage was taken of the highest point of the terrace, and when the streets were cut through it, on its northern and southern ends, the grade was nearly twenty feet lower than the top of the terrace upon which it was erected. A casual observer, therefore, would be likely to take the whole as artificial, whereas more than one-half, as it then appeared, was of fluviatile origin. The dividing line between the natural ground and the mound proper is shown in the engraving. It is about midway between the level of the street and the top of the mound.

The demolition of this ancient landmark was an event which awakened much interest among the citizens, who gathered in crowds, from day to day during the many weeks occupied by its removal. Numerous and conflicting accounts were published at the time concerning it, with any amount of speculation and hasty conclusions. Some of them have been perpetuated in one recent work, at least, upon the pre-historic races of America; on which account I think it proper to say that the statements which follow are based upon personal and careful examination of the work during the process of its removal, until its destruction was accomplished.

This mound, as is well known, was used by the Indians as a burial place, and only about sixty years since, it was visited by a small band, who disinterred and carried away the bones of their chief who had been buried there. But their interments here, as was their unvarying custom, were near the surface. I have observed the same in other localities, sometimes not more than eighteen inches from the top of the mound,—as was the case with some I examined in Washington County, on the banks of the Missouri. On account of this it is not difficult to distinguish the Indian burials from those of the Mound-builders. Had this fact been better understood, we would have been spared many erroneous statements, as well as hasty generalizations upon articles taken from the mounds, which were attributed to their builders, but which, in fact, were deposited by the Indians; and many of them even, subsequent to their first acquaintance with our own race. A striking example of this occurred during the removal of the "Big Mound." Near the northern end, and about three feet from the surface, two skeletons were discovered very near each other, one evidently that of a male, the other a female. With the larger of the two were found the spiral spines of two conch shells, much decayed, nine ivory beads of an average size, as near as I can recollect, one inch in length and nearly one-half in

diameter, an ivory spool with short shaft but very wide flanges, which
were much broken around the edges, and two curious articles of copper,
about three inches in length and about half as wide, resembling somewhat in shape the common smoothing iron of the laundry The under
side, which was concave, showed the marks of the mould in which they
were cast. The upper side, which was much corroded, showed traces
of an elaborate finish in the way of engraving. From the center of the
finished upper side an arm projected at a right angle, about five-eighths
of an inch in continuous width and two-eighths in thickness at its junc-
ture, which tapered to a thin edge.

Embedded in the verdigris with which they were encrusted were plainly
visible the marks of a twisted string just like ordinary wrapping twine,
which had been clumsily tied about them, and upon which the beads had
been strung. All the above articles were about the head and neck of the
skeleton, and had evidently been interred with the possessor just as he
wore them in life.

I have been thus particular in the account of this "big Indian" and his
treasures—for such he undoubtedly was—because these articles of copper,
and the ivory spool, which must have been turned in a lathe, (and I
must include also the pieces of cloth found with them, which however I
did not see) have been taken as the exponents of the state of the arts
among the Mound-builders, and have been made the subject of the most
extravagant statements. Although I was not present when these articles
were taken out, they were placed in my hands a short time afterward,
by the person who unearthed them, who also kindly gave me portions of
the skull, the larger bones of the legs, and a *lock of hair!* from the head
of both the sachem and his squaw, which are still in my possession.

But the most interesting feature of this truly great structure is the
sepulchral chamber which it once contained. By what means the
ponderous mass of earth which formed its roof was sustained, the mound
itself furnished no clue, for it had long ago fallen in and crushed almost
to atoms the already decayed bones of the skeletons lying upon the
floor. The original length of the chamber could only be conjectured, as
portions of the mound had been removed when the street was cut
through upon the southern end, as seen in the engraving. It could be
traced, however, for seventy-two feet. For this distance the sides were
perfectly smooth and straight, and sloped outwardly a few degrees from
the perpendicular, and the marks of the tool by which the walls were
plastered could be plainly seen. One circumstance, which was very
puzzling for a while, was the curious appearance of the surface of the

walls. They were covered with a complete network of black lines, interlacing and crossing each other with all sorts of beautiful and fanciful complications, resembling more than anything else the delicate tracery of a frosted window pane. Upon careful examination, these proved to be the remains of rootlets from the trees which once grew upon the surface above; which, finding easy ingress along the face of the wall, had thus covered its surface, but were now completely carbouized.

The manner of its construction seems to have been thus: The surface of the ground was first made perfectly level and hard; then the walls were raised with an outward inclination, which were also made perfectly compact and solid, and plastered over with moist clay. Over these a roof was formed of heavy timbers, and above all the mound was raised of the desired dimensions.[1] The bodies had all been placed in a direct line, upon the floor of the vault, a few feet apart, and equidistant from each other, with their feet towards the west. These were disclosed, several at a time, as the laborers detached long, vertical sections of earth by the simultaneous use of crowbars inserted at the top. Mingled with the black deposit which enveloped the bones, were beads and shells in prodigious numbers, though in no instance were both deposited with the same individual.

The beads, so called, are the same as are found in the mounds of Ohio, and evidently cut, as Dr. Foster thinks, from the *Busycon*, from the Gulf of Mexico. They are small discs perforated in the center by drilling. From the many specimens in my possession in various stages of their manufacture, the conclusion is warranted that the hole was first drilled and the edges rounded afterwards. Many of these seem to have been cut from the common mussel-shells which are abundant in this region. The small sea shells (*Marginella apicina*), were only found with a few skeletons, possibly five or six at the southern end of the vault, and with each one from four to six quarts, all of which were pierced with small holes near the head, by which they were undoubtedly strung together. With the majority, however, only the perforated buttons were found, but in such numbers that the body from the thighs to the head must have been covered with them.

Being very desirous of securing, if possible, a perfect skull, or at least the fragments from which one might be reconstructed, and as all

---

[1] Although not a vestige of wood was discovered when it was removed, in a work across the river, more recently destroyed, which contained a similar vault, were found sticks of red cedar, much decayed, but in such positions as showed that they had been the supports of the superincumbent earth.

which were thrown out by the excavators were in small pieces which
crumbled at the touch, I began a careful excavation with a common
kitchen knife near the feet of a skeleton, following the spinal column to
the head. My work was soon interrupted however by the crowd of
eager boys from the neighboring schools, who scrambled for the beads
which were thrown out with every handful of earth, with such energy
that I was lifted from my feet and borne away. By the aid of a burly
policeman, however, I was able to finish my excavation, but without
being able to secure what was so much desired. The bones were so much
decayed, when the roof fell in, that all the larger bones were crushed,
and only small fragments of the skull could be obtained, and of course
no cavity corresponding to its shape remained from which a plaster cast
might have been taken.

The last visit to the mound was most interesting of all. The
night before, the workmen had made a vertical cut directly across
the northern end of the small portion of the work which yet remained.

Cross-Section of the Big Mound at St. Louis.

What was there revealed is well represented in the engraving. The
sloping walls were of compact yellow clay, the intermediate space filled
with blue clay in a much looser condition, in perfect agreement with the
idea of its having fallen in from above by the decay of its support. Here
too, at the northern end, I conjectured, was the entrance to the sepul-

cher, for the reason that here the walls were about eight feet in height, from six feet to eight feet apart, whereas the first measurements at the top, when the walls were discovered, showed a diameter of eighteen feet.

Here, then, was an artificial sepulchral tomb, whose dimensions we may safely state to have been from eight to twelve feet wide, seventy-five feet long, and from eight to ten feet in height, in which from twenty to thirty burials had taken place. If any other deposit had been made with the dead, save the before-mentioned beads and shells, the tomb must have been desecrated by some savage who had no regard for its sacred character, for not a vestige of anything else was disclosed at the time of its demolition.

Another evidence of a large aboriginal population is furnished by the stone mounds which are very numerous in certain localities, particularly in those counties through which flow the Osage and the Gasconade rivers. Not being so conspicuous as the others already noticed, they would not be likely to attract the attention of ordinary travelers, and may therefore be found covering a much larger area than is at present known. These are simple heaps of stones, of such size as could be conveniently carried from the ravines where they are found to the highest elevations—the spots usually chosen for their erection. I have seen them in groups on a continuous line running back from the very brow of a precipitous escarpment two hundred and fifty feet above the Gasconade, which swept majestically below. In fact, those commanding elevations, no matter how difficult of access, from whence the view of the surrounding landscape was most extended and lovely, seem to have been the ones most preferred. The Ozark Hills, clothed with the primeval forests, are full of them. They are generally considered more recent than the earthen tumuli. In all that I have opened nothing was discovered which shed any light upon their history, save a few human teeth and the smallest bits of the larger bones, which proved them to be burial mounds. It is stated by Adair that some of the nomadic tribes of Indians thus disposed of their dead, and as they passed and re-passed those graves, from year to year each man of the tribe was accustomed to add another stone to the heap which had been raised above them. In a group of seven, I observed one which showed some skill in masonry; one of the walls was built up with a smooth face about three feet in height, in which the joints were beautifully broken, although there was no evidence of mortar having been used.

In this connection should be noticed still another class; the most noteworthy examples of which, were discovered about the year 1818, in the town of Fenton, about fifteen miles from St. Louis. These were stone graves or cists, each inclosing a single skeleton, or the dust of one—as all were in a crumbling condition there. Not one of the many examined exceeded fifty inches in length. They were built of six flat stones, single slabs forming the bottom, top, sides, and ends.

According to Dr. Beck, of Gazetteer fame, much discussion was elicited at the time and many communications appeared in the newspapers. The chief point upon which it all centered was the shortness of the graves. As was the case in Tennessee, a few years since, it was considered as proving the former existence of a race of pigmies. But the fact that in some of them the leg bones were observed lying parallel with and along-side of the bones of the thigh, accounted for the shortness of the graves; and this, taken along with the well-known custom practiced by some tribes, of suspending their dead in the branches of trees until the bones were denuded of flesh and afterwards depositing them in their common burying place, was regarded as a sufficient answer to all the pigmy speculations.

About one hundred yards from the ancient burying ground at Fenton were once a number of mounds, and remains of an extensive fortification, which also attracted the attention of the curious in those early days. And if files of the old *Missouri Gazette* of sixty years ago could be found, no doubt many interesting facts would be recovered which are now forgotten. Similar stone graves are found in Perry County, seventy-five miles from St. Louis.

CHAPTER VI.

To the general student of Ethnology and Archæology, no one department of antiquarian research has yielded grander or more satisfactory results than those which have rewarded the explorers of the caves and rock-shelters of some of the mountain chains of the old world. Concerning the relative age of the earthen structures of the vast alluvial plains of America there may be much difference of opinion. But in his occupancy of the caves of Europe, primeval man has so inscribed the records of his early life and presence, during those geologic changes which he witnessed, in the succession of the glacial and diluvial epochs, that they are sometimes as sharply delineated and legible as are those of the various orders of animal life in the stratified rocks. By these faithful chronographs of the childhood of the race, we are carried back irresistibly to a period so remote, that the cave-dwellers from Mount Hor, who joined the confederate kings, and were so signally overthrown by Abram in the plains of Sodom, were but of yesterday.

In America, this field is comparatively unexplored, or perhaps we had better say, is undiscovered. Indeed, it may be that we have nothing here which shall be found to correspond to or compare with the drift period and bone-caves of Europe. It is true we find, in the early tales of border life in Virginia, Kentucky and Tennessee, accounts which must contain some elements of truth, of caverns filled with human bones; others whose walls are pictured and sculptured with strange devices, of animals, known and unknown; and representations of the heavenly bodies; and others still, containing mummied corpses, embalmed and wonderfully preserved, clad in robes of feather-work like those of Peruvian fabric which so filled the Spanish conquerors with admiration. But alas! these were long since destroyed. Then, they had little or no scientific value, consequently there was no motive for their serious examination, or preservation.

Still, however, we may indulge the not unreasonable hope that others may yet be discovered, whose disclosures shall be equally precious. In this hope we are the more encouraged by the fact that the few which have been noticed and described, furnish indubitable proof that they

were once the favorite resorts, for burial purposes, of some pre-historic
race.   When the stones shall be rolled away from the doors of the sepulchral caverns in the limestone hills of Missouri, the long-forgotten
dead may again come forth re-vivified, rehabilitated, and the Ozark
Mountains may yet disclose materials for a chapter in the life of her
primitive people, which shall equal in interest the records of the mounds.
The Ozarks, thanks to their sterile slopes, have preserved their sacred

Among the Ozarks.

treasures well.   They are honeycombed with caves, some of unknown
extent.   Their openings may be seen in the precipitous bluffs along the
Gasconade River, in great numbers, on either side, or the majestic arches
of their openings span the divides where the smaller hill ranges meet.
Do these numerous caves and channels evidence an ancient system of
drainage, in operation long before the Gasconade had asserted its "right
of way" and scooped for itself a course through the rocks by its ceaseless flow? [1]

In these caves the ancient dead were buried and the funeral feasts
were celebrated.   The deep deposit of rich nitrogenous earth in the

[1] See Sir Charles Lyell's remarks upon the Valley of the Meuse, "Antiquity of Man,"
p. 73.

larger chambers, and the bones of various animals, birds, and mussel
shells—the refuse of the funeral feasts,—the alternate layers of ashes and
charcoal mingled with earthy matter, containing human bones in different
degrees of preservation, tell of oft-repeated visits and recurrence of the
funeral rites.

What little we have learned from the few thus far explored makes us
only the more eager to examine still further the records they contain.
A description of one must serve our present purpose. The one selected
is in Pulaski County, and is one of the many famous saltpetre caves so
often mentioned in the early annals of the State, with which the country
of the Gasconade abounds. The opening is in the face of a perpendicular limestone bluff which extends along the river for many miles.
While the scenery of this whole region is very beautiful, the view from
the mouth of some of the caves is enchanting. Standing in the shadow
of one of their lofty arches, the eye is charmed with the peculiar beauty
of the landscape spread out before it. The Gasconade flowing far below,
the stately trees which fringe its banks and mark the course of its long
graceful curves, until it loses itself in the dim outlines of the Ozarks,
which swell and roll away until their opalescent hues melt into the mel-
low light of the autumn sky,—all conspire to awaken the liveliest
feelings of respect and admiration for a people whose æsthetic taste was
so refined and tender as to lead them to select a place so charming for
the long repose of their loved ones. But poetry and science have but
little in common: one must end where the other begins. So turning my
back upon the beautiful scene, and repressing all compunctions for the
sacrilege we are about to commit, the impatient workmen are directed
to begin the labor of cutting a trench one hundred and seventy-five feet
long, through the deposit at the bottom of the cave. At the end of this
distance the perpetual gloom begins. Here the torches are brought into
requisition, by whose dim light, as the laborers proceed with their work,
the sectional notes and measurements are taken.

The whole surface of the deposit seems to have been much disturbed,
to the depth of from eighteen inches to two feet. It is composed of
earth and ashes, mingled profusely with broken pottery, fragments
of human bones and flint-chips. Below this, the deposit is hard and
compact. Selecting a point about midway from either end of the trench,
we proceed to make more critical examination. Continuing the excavation to the depth of six feet, the natural deposit at the bottom is
reached, composed of a tough reddish clay, which contained nothing
but decayed mussel shells. All above this showed the continual

4

occupancy of the cave during its deposition. A vertical section at the point above named, disclosed the following strata:

Alluvium, mingled with ashes, bits of pottery, etc.	18 inches.
Stratum of different colored ashes	2 "
Clay and dark Alluvium	2½ "
Ashes	½ "
Alluvium	3 "
Mixture of Ashes and Clay	3 "
Pure Ashes	½ "
Alluvium	3½ "
Pure Ashes, mingled with Charcoal	4 "
Alluvium, " " "	7 "
Ashes	3 "
Alluvium, mingled with Charcoal	20 "

At the depth of about two feet, the first skeleton was reached, lying upon its back, with head towards the east. All the small bones were thoroughly decayed. About six feet north of this, another skeleton was disclosed, evidently buried in a sitting posture. This was so much decomposed that only a few of the thicker portions of the skull could be secured. Near this was also found the skeleton of a very aged female, the skull in a better state of preservation. In companionship with these was a flint spear-head of the rudest pattern, as were all the implements of stone—which were not numerous—which the deposit contained. With the exception of the rude spear-head, their presence seemed to have been accidental, and this also may have been so. Among the most interesting relics, were articles of bone, such as awls, scrapers, and the like, and occasionally one made from the inner surface of a shell, with a sharp edge.

What was most surprising was the prodigious number of mussel shells which were continuous through the whole deposit, decreasing in size and more decayed as we descended, until their whole substance was a chalky paste. These are still abundant in the river below. Intermingled with the alluvium and ashes, as far as the excavation extended, were skulls and bones of fishes, deer, bear, mud-turtle and wild turkey. The skulls were always broken, no doubt to obtain the brains, which have always been esteemed a great delicacy among the civilized and savage as well. While, for purposes of ethnological study, a more detailed description of the crania contained in this cave would be instructive, and other particulars here suggested might be properly enlarged upon, still, enough has been stated to indicate the desirableness of a more thorough exploration of this comparatively new class

f antiquities. But keeping in mind that we have more to do in these
chapters with the traces of the aboriginal inhabitants of Missouri than
with lengthy generalizations upon the facts they disclose, we can only
hint at one or two conclusions.

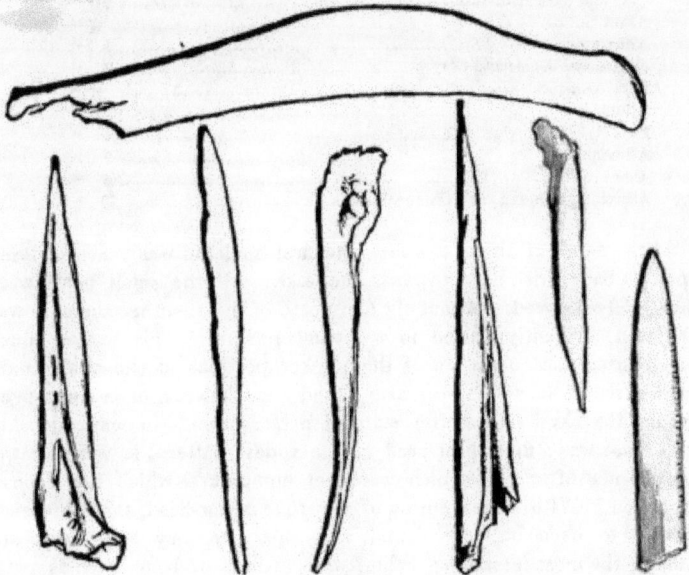

Bone Implements.

Here was the burial place of a people who were not insensible to
those beauties with which nature around them was glorified, and who
sought those places with the most lovely surroundings in which to deposit
the remains of their friends. Here were laid to rest from time to time
the old and young, the aged matron, and the child, the fragments of
whose thin, paper-like skulls suggested many thoughts of maternal love
and tears of sorrow. The vast numbers of shells, and bones of beasts
and birds, bear witness to the oft-repeated funeral feasts beside the new-
made graves of the departed, and point to a belief in a life continued in
another world. Who they were, or when they lived, it is not our
province now to try to answer. The Indians, it is well known, regarded
these gloomy caverns with superstitious fear, for in them they believed

the great Manitou dwelt. In view of this fact, so well attested by early
writers, the idea that they were the occupants becomes a matter of grave
doubt. The skulls thus far examined, are also wanting in those peculiar
and generally very marked characteristics which are so evident in the
crania of the mounds. With this allusion to a question so interesting,
we must leave its discussion to a future occasion, when we may reasonably hope to be able to continue it in the light of more extended
information.

CHAPTER VII.

Although the propriety of some of the mound-classifications of the earlier writers has sometimes been questioned, no doubts are entertained as to the purpose of those which have been denominated Temple Mounds. In treating of this class, we enter at once upon a field almost as vast as the two continents of America. For, whatever may have been the material used in their construction, whether stone, or earth alone, or both combined, they present such uniform characteristics, so identical in evident purpose and design, that they link together by one prevailing system of religious worship, of which they are the striking exponents, unnumbered tribes and peoples, scattered up and down the two continents from the Atlantic to the Pacific. Reason as we may, the more they are studied, and considered in their relation to other groups and classes with which they are found associated, we can hardly escape the conviction that they point to one common origin.

Before yielding a hasty assent to a general conclusion, a proper caution would suggest the possibility of accounting for this uniformity of structure by other and natural considerations. It is well known that barbaric tribes in all lands and times have manufactured their first implements of war and the chase from stone and bone, and have learned, by means of some hint which Nature, perhaps, afforded, to fashion rude vessels of clay for domestic use. It is also true that their petitions and adorations have been addressed to the same class of imaginary beings, or objects and active forces whose effects they were accustomed to behold around them; among which the heavenly bodies appear to have occupied a conspicuous place, particularly during some stages of their progress from barbarism to a higher life.

Possessed of the same faculties, appetites and passions, inheriting the same necessities, meeting always the same difficulties in their struggles for existence, it is not surprising that rude nations have ever followed the same paths in all the activities of their wild, infantile life. Indeed, it would be surprising if they had not. From these and similar considerations it may be thought that the identity of form, structure and relation, and also apparent oneness of purpose which characterize the Temple Mounds, demonstrate only the operation of a universal law, in the progress of a people from a state of barbarism through the slow stages of its developement towards a higher civilization. The sun and moon have been worshiped in ages and countries widely separated, and by nations between which there could never have occurred any possible communication.

Man never has attained by intuition or philosophy that knowledge of the unity and perfections of the Supreme Being which Revelation presents : and wanting that knowledge, he naturally worships those visible objects which are most conspicuous and which most inspire his reverence, especially those which, he conceives, exert the greatest influence upon his life and destiny. But when each nation starts out for itself in the path of a progressive civilization, the prevailing forms of worship, being subjected to the same influences which mould the national polity, must necessarily, under the new impulse, become also materially changed, or as has sometimes happened, displaced altogether, by a system entirely new. From this point, the forms of Nature-worship would cease to be identical, and each resultant system become thereafter more and more divergent ; and long periods of time must necessarily be required for the working out of a complicated and well arranged system of popular religion which should be able to enforce the ready obedience and subjection of a vast people to its mandates, and enlist the energies of the nation in the erection of their most imposing structures, for no other purpose than the observance of their religious rites and ceremonies. Such structures, among the memorials of an ancient people, are very interesting and instructive, from the fact that religion has ever exerted such controlling influence in the establishment and perpetuity or decline of countless nations, whose history has been preserved.

They are the records, therefore, of more than the religious faith and practice of a particular people ; but, because of the leavening influence of religious ideas when crystallized into systematic forms, they become the interpreters of many things which otherwise would never be understood.

It will readily be seen, therefore, in the light of the foregoing, that the Temple Mounds of America are invested with an interest and importance outside of their purely religious character ; and which is greatly enhanced by the fact that wherever they are found, along with them invariably occur the most striking evidences of the former presence of a numerous population, whose civil and social condition was separated by a wide gulf from that of the red race who occupied their ancient seats when America was discovered; and whose government was so well established and enduring, as to render it possible for vast numbers to be employed for a series of years in their erection.

Temple Mounds, according to Squier and Davis, "are distinguished by their great regularity of form and general large dimensions. They consist chiefly of pyramidal structures, truncated, and generally have graded avenues to their tops. In some instances they are terraced or have successive stages. But whatever their form, whether round, oval, octangular, square or oblong, they have invariably flat or level tops."[1]

"The summits of these structures were probably crowned with temples, but having been constructed of perishable materials, all traces of their existence have disappeared. The truncated pyramidal form, which often rises to no great height, was obviously the foundation for such structures. In the works at Aztalan, Wisconsin, we trace the outlines of this form of mounds at the angles of the bastions, and this may be said to be their northern limit. They are not recognized on the southern slope of Lake Erie, and are seen at only three points in Southern Ohio, viz: Marietta, Newark and Chillicothe.

"The stupendous mound at Cahokia in Illinois, with its graded way, its terrace and level summit, was the best representative of this class.[2] In Kentucky they are not rare; the great mound near Florence is of this character, and that near Claiborne—fifty feet in height—has a level summit with a gradual slope on the east, and a succession of ten terraces on the west. In this class, too, must be included the great mound at Seltzertown, Mississippi, and most of those in the Gulf States.

---

[1] Ancient Monuments of the Mississippi Valley, p. 173. Smithsonian Contributions to Knowledge.

[2] When he wrote this, Dr. Foster was under the impression that this great work was destroyed. While he was mistaken, it is understood to be for sale, and may soon be reckoned among the things that were, provided some railway shall be constructed near enough to render its huge mass—containing several million cubic yards of earth — desirable to elevate its grade. What a graceful thing it would be for the State, or National Government to purchase it and decree its perpetual preservation! Men of science all over the world and in all future time would be so thankful for such an act. Thus the Government of Denmark has done with her antiquities. Whether either of our great political parties could be persuaded to assume such a tremendous responsibility is very doubtful. Our legislators are so conscientious and so intent upon "retrenchment and reform," that the expenditure of a few hundred dollars for the preservation of the stupendous work which must have occupied the ceaseless labor of thousands of men through a life-time to erect, would be a precedent too dangerous to think of—such an act might shake the foundations of the Republic. No partizan would dare favor such a proposition, lest it should be followed by his speedy consignment to a political grave from which there could be no resurrection.

"In Mexico and Central America, we see the culmination of this form in the Teocallis, which were faced with flights of steps and surmounted by temples of stone." [1]

The identification of some of the mounds in their enumeration as Temple Mounds, by the authors above named, I cannot but regard as lacking confirmation. Indeed the evidences derived from my own observations are conclusive that some of them belong to quite another class. Those of the "truncated pyramidal form which often rise to no great height," were doubtless crowned with the residences of the chiefs and rulers. These are often found in groups. I have counted seven or eight very near each other, a few feet in height, with flat or level tops: the central one generally larger than those around it, which tradition affirms was occupied by the dwelling of the chief. The others of the group were erected from time to time for residence sites for his sons, as they came to man's estate and had families of their own. In all which I have excavated, nothing was disclosed but fragments of pottery.

The only structures which can with certainty be identified as Temple Mounds are those whose perfect model is seen in the Teocallis of Mexico and South America.

In whatever group they are found, they are the most imposing. Generally oblong, with one or more stages, and ascended by graded avenues. Such was one of the large mounds at St. Louis, and I am disposed to believe that the beautiful Falling Garden was an unfinished work of this class, whose three stages, about fifteen feet each in height, were finished, but the elevated work which was to crown the whole was wanting.

The great Cahokia Mound is the best representative of this class to be found in North America. This was examined by Mr. Brackenridge in 1811-12. His interesting description of it, along with the numerous works of smaller dimensions with which the American Bottom is filled,— or was in his day—may well be quoted entire in this connection:

"To form a more correct idea of these, it will be necessary to give the reader some view of the tract of country in which they are situated. The American Bottom is a tract of rich alluvial land, extending on the Mississippi, from Kaskaskia to the Cahokia River, about eighty miles in length and five in breadth; several handsome streams meander through it; the soil is of the richest kind, and but little subject to the effects of the Mississippi floods. A number of lakes are interspersed through it, with

---

[1] Foster's Pre-Historic Races, etc., p. 186.

fine high banks; these abound in fish, and in autumn are visited by
millions of wild fowl.

"There is perhaps no spot in the western country, capable of being
more highly cultivated, or of giving support to a more numerous
population, than this valley. If any vestige of ancient population were
to be found, this would be the place to search for it; accordingly this
tract, as also the bank of the river on the western side, exhibits proofs
of an immense population. If the city of Philadelphia and its environs
were deserted, there would not be more numerous traces of human
existence.

"The great number of mounds, and the astonishing quantity of human
bones everywhere dug up, or found on the surface of the ground with
a thousand other appearances, announce that this valley was at one
period filled with habitations and villages. The whole face of the bluff, or
hill which bounds it on the east, appears to have been a continued burying
ground. But the most remarkable appearances are two groups of mounds
or pyramids, the one about ten miles above Cahokia, and the other
nearly the same distance below it, which in all exceed one hundred and fifty
of various sizes. The western side also contains a considerable number.

"A more minute description of those above Cahokia, which I visited in
the fall of 1811, will give a tolerable idea of them all. I crossed the
Mississippi at St. Louis, and after passing through the wood which
borders the river, about a half a mile in width, entered on an extensive
plain.

"In fifteen minutes I found myself in the midst of a group of mounds,
mostly of a circular shape and at a distance, resembling enormous haystacks
scattered through a meadow: one of the largest, which I ascended, was
about two hundred paces in circumference at the bottom, the form nearly
square, though it had evidently undergone considerable alteration from
the washing of rains; the top was level, with an area sufficient to contain
several hundred men. The prospect from this mound was very beautiful;
looking towards the bluffs, which are dimly seen at the distance of six
or eight miles, the bottom at this place being very wide, I had a level
plain before me, bound by islets of wood, and a few solitary trees: to
the right the prairie is bounded by the horizon; to the left, the course
of the Cahokia may be distinguished by the margin of wood upon its
banks, and crossing the valley diagonally S. S. W. Around me, I
counted forty-five mounds or pyramids, besides a great number of small
artificial elevations: these mounds form something more than a semi-
circle, about a mile in extent, the open space on the river.

"Pursuing my walk along the bank of the Cahokia, I passed eight others in the distance of three miles, before I arrived at the largest assemblage. When I reached the foot of the principal mound, I was struck with a degree of astonishment not unlike that which is experienced in contemplating the Egyptian Pyramids. What a stupendous pile of earth! To heap up such a mass must have required years, and the labor of thousands. It stands immediately on the bank of the Cahokia, and on the side next it, is covered with lofty trees. Were it not for the regularity and design which it manifests, the circumstances of its being on alluvial ground, and the other mounds scattered around it, we would scarcely believe it the work of human hands.

"The shape is that of a parallelogram standing from north to south; on the south side there is a broad apron or step, about half way down, and from this another projection into the plain, about fifteen feet wide, which was probably intended as an ascent to the mound. By stepping around the base I computed the circumference to be at least eight hundred yards, and the height of the mound about ninety feet. The step or apron has been used as a kitchen-garden by the monks of La Trappe, settled near this, and the top is sowed with wheat. Nearly west there is another of a smaller size, and forty others scattered through the plain. Two are also seen on the bluff at the distance of three miles. Several of these mounds are almost conical. As the sward had been burnt, the earth was perfectly naked, and I could trace with ease any unevenness of surface, so as to discover whether it was artificial or accidental.

"I everywhere observed a great number of small elevations of earth to the height of a few feet, at regular distances from each other, and which appeared to observe some order; near them I also observed pieces of flint and fragments of earthen vessels. I concluded that a populous town had once existed here, similar to those of Mexico described by the first conquerors. The mounds were sites of temples or monuments to great men.

"It is evident this could never have been the work of thinly-scattered tribes. If the human species had at any time been permitted in this country to have increased freely, and there is every probability of the fact, it must, as in Mexico, have become astonishingly numerous. The same space of ground would have sufficed to maintain fifty times the number of the present inhabitants, with ease, their agriculture having no other object than mere sustenance. Among a numerous population, the power of the chief must necessarily be more absolute, and where there are no laws, degenerates into despotism. This was the case in Mexico, and in the nations of South America. A great number of individuals were at the disposal of the chief, who treated them little better than slaves. The smaller the society, the greater the consequence of each individual. Hence, there would not be wanting a sufficient number of hands to erect mounds or pyramids."

The largest mound of the Cahokia group, thus described by Brackenridge, is now known as Monks' Mound, on account of its having been occupied in early days by a colony of monks of the order of La Trappe. This prodigious temple site, as before remarked, is the best representative of its class in the United States, not only on account of its vast size, but also because it is the most finished model of all similar works which can with any degree of certainty be determined as temple mounds. The Teocallis of Mexico and the regions further south, though finished with stone, are of the same form, with graded ascents, or flights of steps, leading to the broad stage, or level top, at one end of which rose another elevation, upon which stood the most holy temple and sacred altars.

Upon these burned the perpetual fire, to be extinguished only at the close of the year, and rekindled by the sun himself, as his rising beams were concentrated by the high priests, when the new year began. This event was always observed with the greatest solemnity.

When the sacred flame expired upon the altars, with the dying year, the whole land was filled with gloom, and the fire upon every domestic hearth must be extinguished also. Then the people sat down in awful suspense to watch for the morning. Possibly their father, the sun, might be angry with his children, and veil his glory behind the clouds at the coming dawn. Then as they thought of their sins and bewailed their transgressions, their fears were expressed in loud lamentations. But as the expected dawn—the momentous time—approaches, all eyes are turned towards the holy mount where the now fireless altars stand. At length the eastern sky begins to glow with a golden light which tells them that their god is near, and, while they watch, he rolls in splendor from behind the eastern hills, and darts his fiery beams upon the sacred place where holy men are waiting to ignite anew the sacrificial fires. Nor do they wait in vain, for soon the curling smoke and the signal flames are seen by the breathless multitude which fill the plains below, and then one long, glad shout is heard, and songs of joy salute the bright new year. Swift-footed messengers receive the new-lit fire from

the hands of the priests, quickly it is distributed to the waiting throng
and carried exultingly to their several homes, when all begin the joyful
celebration of the feast of the Sun.

The peaceful tribes who once dwelt in this region of the Mississippi
Valley, upon either shore, found no quarries of stone of easy cleavage,
or which could be wrought with their simple tools for the erection of
their edifices. Doubtless wood was the only material at their command,
or possibly sun-dried brick. The dust of their temples is gone with that
of their builders; their altars are crumbled—the sacred fire is extinguished, which the sun shall nevermore rekindle. But the proud monument of their national solemnities still rears aloft its majestic form in the
midst of a vast alluvial plain of exhaustless fertility—a grand memorial
of days more ancient than the last migration of the Aztec race to the
plains of Anahuac, who found there the very same structures, which
they appropriated and by which they perpetuated the worship of the
land of their fathers as well as that of the people whom they subjugated.
It is not unreasonable to suppose that when, from its elevated summit,
the smoke of the yearly sacrifice ascended in one vast column heaven-
ward, from the great work above described, that it was the signal for
simultaneous sacrifices from lesser altars throughout the whole length of
the great plain, in the centre of which it stands, and that the people
upon the Missouri shore responded with answering fires from those
high places which once stood upon the western bank of the river, but are
now destroyed.

Here, we may well believe was the holy city, to which the tribes made
annual pilgrimages to celebrate the national feasts and sacrifices. But
not here alone ; for this vast homogeneous race, one in arts and worship,
had the same high and holy places, though of less imposing magnitude,
in the valley of the Ohio, in Alabama, and Mississippi.

In south-east Missouri, at New Madrid, is a similar work, surrounded
by a ditch ten feet in width and five in depth. It is twelve hundred feet
in circumference and forty feet in height. Among the ruins of almost
every ancient town lying back from the river, upon bayous and smaller
streams, may be found the oblong Temple-mound, which is always the
highest work of the group, and commands a view of the whole.

There are some who profess to believe that the Indians are the degenerate sons of the authors of these extensive and complicated works. But
when it is remembered that their languages, which are divided into many
groups, present very few affinities which are common to all, and the
dialects into which these groups are further divided are, many of them,

so distantly related as to show that the various tribes must have been separated from the parent stock in times very remote; and when we take into the account also, the wonderful unity of the race of the mounds, as displayed in their works and worship, and the vast extent of territory they occupied, it will be seen that such a supposition involves an antiquity of the red race, which its most ardent defenders will find difficult to harmonize with the recognized facts.

To my own mind the evidence is clear that the two peoples were as distinct as the Greeks and Romans. That the exodus of the mound-builders occupied long periods of time, is altogether probable, and comprised several distinct migrations, to the south and southwest, which were brought about by the continued encroachments of the more warlike and savage hordes from the north and northwest. Here and there, no doubt, small bands were enslaved or absorbed by their conquerors, who adopted some of the customs of the subjugated race, particularly those pertaining to their worship, the traces of which are often well defined,—the practice of which was continued by a few Indian tribes as late as the beginning of the present century.

If the views here presented are correct, it will be apparent that the Temple-mounds are invested with an interest peculiar to themselves, in as much as they give us an insight to the social and political condition of the ancient inhabitants of the State of Missouri and the Mississippi Valley, which can be gained from no other class of works. It will also be perceived that we have barely entered upon a most interesting field of research, which will well repay a careful and thorough examination.

CHAPTER VIII.

The foregoing evidences of an ancient people swarming in prodigious
numbers throughout the vast territory in which these works abound, and
who had their permanent dwellings in towns and cities which were well
arranged and constructed with no mean skill, suggest the most interesting question, How did they subsist? The importance of this question is
realized when we remember that it lies at the foundation of their whole
social fabric; and in fact, once determined, the answer becomes one of
the chief exponents of their physical condition, intellectual capacity and,
in a good degree, of their moral status as well. Many of the staple articles of food upon which all civilized nations depend for subsistence are
only to be procured by intelligent labor, guided by a plan and forethought
which are the result of a more or less extended observation of nature's
laws.

Here were large cities; then here also must have been trade and commerce of some sort. Merchandise may not have been bartered for gold
and silver, but more likely—as was the case with the Peruvians—the
products of the field, the fold, or the chase, were exchanged for those
of the workshop and domestic handicraft. Again: their means of support must have been so certain and reliable, and withal so abundant, that
large numbers of the people could be employed continuously upon those
monuments of their industry which they have left behind for our admiration. The probability that fish formed no inconsiderable item of their
food supply has already been suggested. The name of our great river,
which it is thought has come down to us from their time—*Nemesi-sipu,*
which means River of Fish—if it be true, bears witness to this. The
prodigious shell heaps along the southern coast, from Florida to the
mouth of the Mississippi, may also be noticed as evidence of the fact
that they were not unskillful fishermen. These accumulations of
the refuse of their kitchens have often proved peculiarly interesting
and instructive, inasmuch as they abound in numerous relics which,

under other circumstances would have been destroyed. The shell heaps of the Baltic coast are complete zoological museums of the fauna of the period when they were formed, containing, as they do, the bones of many animals long since extinct in those regions, and presenting also the bones of the few domestic animals which were the companions of man in that remote period.

The most important sites of the towns of the pre-historic Americans are found upon the shores of lakes or banks of rivers, and generally—though not always—contiguous to, or upon extensive areas of fertile land. We are not compelled to suppose, however, that they were always influenced by agricultural considerations in the location of their permanent homes, for the ruins of some towns have been observed upon the sandy beaches of lakes, and where, too, there was no fertile land near, which was suitable for agricultural purposes. It is, therefore, natural to suppose that the inhabitants of towns so situated were fishermen.

The wide, deep ditches on the inside walls of some of their enclosures have called forth much speculation as to their purpose. It has generally been assumed that the walls which enclosed their towns were erected for defensive purposes. But the puzzle has been about the location of the ditch along the base of the wall within the enclosures. According to all our notions of warfare, the ditch—to serve any defensive purpose—should have been outside the walls. Moreover, many of the walled towns were so situated in valleys, which were overlooked by the near hills which surrounded them, as to be totally incapable of defense in any kind of known warfare. The theory, therefore, that this inside ditch was one of their means of defense, seems hardly satisfactory. I have somewhere met with the statement that there was a tradition to the effect that the ditches were receptacles for water, or rather, artificial channels for water conducted from the natural streams near which the towns were located, thereby furnishing the inhabitants with a constant and flowing supply. Without stopping to discuss the question, it may be remarked that the idea seems altogether probable, and their construction for such a purpose a very natural thing to do, while the control of the stream, by gates and locks, would require no greater engineering skill than they have displayed in their more durable works. They would also have been specially adapted to the culture of fish, or they may have been the receptacles for their winter's supply. Speculative as the above may appear, it is certainly as rational as the notion that the inside ditch contributed in any way to their defense against the attacks of their foes.

What sort of domestic animals, if any, were reared by the ancient inhabitants of Missouri, we have no knowledge; but there can be very little doubt that game was abundant and that they were successful in the chase. There is satisfactory evidence that the huge Mastodon was their cotemporary whose bones are so abundant in our alluvial plains; and also that he was conquered and slain by their seemingly feeble weapons. I have myself exhumed from the ruins of one of those towns fragments of the vertebral column of the buffalo.

However all this may have been, concerning their *agricultural* skill, we are not left to conjecture; and we may confidently assert that their main dependence for subsistence was upon the labors of the husbandman. They worshipped the sun, and invoked his benign influence upon the occurrence of the great annual festival when their crops were sown in the spring; and when these were gathered, in the autumn they offered up the first fruits to him as lord of the harvest.

That this was their custom we may with confidence assume; nor is it, indeed, mere assumption. The largest of these structures—the Temple Mounds—are found to be precisely similar in form and character to those of Mexico; and the Spanish historians have given the fullest accounts of the manner in which their religious exercises were performed upon their summits, or in the temples which crowned the Teocallis. And as the belief prevails that the builders of these were of the same race as the Moundbuilders, and probably their descendants, it becomes almost certain that structures of the same form in both countries were erected for the same uses and ceremonies. If it be true, as we believe, that when the great majority of the race of the Mound-builders had been destroyed, or driven from their habitations in the Mississippi Valley, some of whom are known to have migrated to the southwest—some remnants of the tribes remained, and were absorbed by their conquering successors, then we might expect to find some of the customs of their fathers still practiced by those who were left behind; and more particularly, those pertaining to their religious rites and manner of providing for their subsistence. The student of the history of the red men cannot fail to notice the fact that a few of the southern tribes possessed traits and customs peculiar to themselves, and in which they differed widely from those of the north and east. The former had a complicated and well-arranged system of religious worship, with the perpetual fire of the altars; also a line of priests or prophets, who enjoined seasons of rigorous fasting, and conducted the exercises upon the occasions of their festivities. The former can scarcely be said to have had any religious system or belief. Mr.

Adair has given a detailed account of the religious rites and ceremonies which were once practiced by a few southern tribes among whom he resided for many years ; and so impressed was he with their imposing and multifarious ceremonials that he believed they must have derived their system from the Jews.

The dissimilarity between the tribes of the south and those of other localities was equally striking in their manner of house-building, sports and games. The former had fixed habitations, in towns with streets and public squares, and a love of home, with various other characteristics which belong to a higher civilization than the nomadic tribes of red men ever possessed.

But perhaps in no one thing was the dissimilarity more strongly expressed than in the methods of agriculture. The author quoted above speaks of having seen deserted cornfields seven miles in extent, and we know that they raised quite a variety of crops, and in abundance, chief among which was maize. Among the now numerous and roving tribes we discover only a methodless and scanty agriculture.

The ancient garden beds supposed to belong to the Mound-builders, which in some instances are several hundred acres in extent, have frequently been noticed in several of the Western States. These are said to have been laid out in straight parallel rows or drills across the fields ; but as none have been found in Missouri, as far as I am informed, they need not be dwelt upon in this connection.

There are evidences of tilling the soil, of quite a novel character, which still exist in prodigious numbers, not only in Missouri but also in other regions west of the Mississippi. I have heard of very few east of that river. These works consist of low circular elevations, generally two or three feet above the level of the natural surface of the soil, with diameters varying from ten to sixty feet ; all are round, or nearly so, sloping off gently around the edges. All that I have seen among the Ozark hills are composed of black alluvial soil, and disclosed, when excavated, no implement or relic of any sort. Their presence may always be detected in cultivated fields when covered with growing crops, by the more luxuriant growth and deeper green of the vegetation. They abound in all the little valleys among the flinty hills of the Ozarks, from Pulaski County, Missouri, to the Gulf of Mexico, and westward to the Colorado in Texas, and as far north as Iowa. Their size in the hilly regions seems to have been determined by the amount of rich vegetable mold which could be scraped together in a given spot. Residence sites they could not have been, or they would have contained some relic of stone or bone,

5

or fragment of pottery, or at least the ashes of the family fire. To enable the reader to form some idea of their prodigious numbers, I can do no better than give the remarks of Prof. Forshey, as quoted by Dr. Foster, in his "Pre-historic Races of the United States," which I take it, refer to the same class.

Says Prof. Forshey: "In my geological reconnoisance of Louisiana, in 1841-2, I made a pretty thorough report upon them. I afterwards gave a verbal description of their extent and character before the New Orleans Academy of Sciences. These mounds lack every evidence of artificial construction, based on other human vestiges. They are nearly all round, none angular, and have an elevation hemispheroidal, of one foot to five feet, and a diameter from thirty feet to one hundred and forty feet. They are numbered by millions. In many places in the pine forests, they are to be seen nearly tangent to each other as far as the eye can reach, thousands being visible from an elevation of a few feet. On the gulf marsh margin, from the Vermillion to the Colorado, they appear barely visible, often flowing into one another, and only elevated a few inches above the common level. A few miles interior they rise to two and even four feet in height. The largest I ever saw were perhaps one hundred and forty feet in diameter and five feet high. These were in Western Louisiana; some had abrupt sides, though they are nearly all of gentle slopes." He further states that he "encountered hundreds of these mounds between Galveston and Houston, and between the Red river and Ouicita; that they were so numerous as to forbid the supposition of their having been the foundations of human habitations; that the burrowing animals common to the region piled up no such heaps; and finally, that the winds, while capable of accumulating loose materials, never distribute them in the manner above mentioned." In conclusion he adds: "In utter desperation I cease to trouble myself about their origin and call them inexplicable mounds."[1]

Fom all that can be learned about them, I see no reason to doubt that they were erected for agricultural purposes, and have therefore presumed to name them Garden Mounds.

It would seem perfectly natural, in a sterile country, and where the inhabitants had few materials for artificial fertilization, to gather into

---

[1] The Professor adds, that " there is ample testimony that the pine trees of the present forests ante-date these mounds." What the testimony is he does not say. If they are the work of the Indians, then we must believe them to have been vastly more numerous than any other facts hitherto known would lead us to suppose.

heaps the thin vegetable mold upon the surface, thus increasing its richness and capacity for retaining moisture. But the question may be asked, why should the same practice be necessary in the prairies and bottom lands, the richness of which is proverbial and inexhaustible. For the answer, we are not left to conjecture

In the rich lowlands of the west, the chief difficulty is too much moisture, especially in seasons of unusual rain-fall. This, the corn-raisers in the American bottom know from repeated experience. Hence, acres of corn are often utterly ruined in such seasons, when planted upon low and level fields which have not ample artificial or natural drainage: when, had the earth been raised a few inches even in drills or mounds, such as have been described, a good crop would have been secured. An intelligent Iowa planter informed me that he had often seen this demonstrated in corn-fields which were filled with these mounds. The low ground between them, if the season were unusually rainy, would yield no returns, while upon the mounds themselves the crop would be excellent. From these considerations, there can be but little doubt that the garden mounds were raised for the better cultivation of maize, which was doubtless the staple article of ancient husbandry. But we are not to suppose, however, that this was the only kind of grain known to the pre-historic Americans; for evidence is not wanting that, in some sections at least, they cultivated wheat, and deposited it, along with those articles which were deemed most precious, in the tombs of their loved ones. Thus—thanks to their affectionate care in the disposition of the dead,—it has been preserved for hundreds, perhaps thousands, of years; and, like the few small grains in the hand of the Egyptian mummy, when brought forth to the sunlight and moisture, has germinated and ripened, and furnished us with a variety unknown before.

From an interesting account of certain mounds in Utah, communicated by Mr. Amasa Potter to the *Eureka Sentinel*, of Nevada, as copied by *The Western Review of Science and Industry*, I make the following extracts:

"The mounds are situated on what is known as the Payson Farm, and are six in number, covering about twenty acres of ground. They are from ten to eighteen feet in height, and from 500 to 1,000 feet in circumference." "The explorations divulged no hidden treasure so far, but have proved to us that there once undoubtedly existed here a more enlightened race of human beings than that of the Indian who inhabited this country, and whose records have been traced back hundreds of years." "While engaged in excavating one of the larger mounds, we

discovered the feet of a large skeleton, and carefully removing the hardened earth in which it was embedded, we succeeded in unearthing a large
skeleton without injury. The human framework measured six feet, six
inches in length, and from appearances it was undoubtedly that of a male.
In the right hand was a large iron or steel weapon, which had been buried
with the body, but which crumbled to pieces on handling. Near the
skeleton we also found pieces of cedar wood, cut in various fantastic
shapes, and in a state of perfect preservation; the carving showing that
the people of this unknown race were acquainted with the use of edged
tools. We also found a large stone pipe, the stem of which was inserted
between the teeth of the skeleton. The bowl of the pipe weighs five
ounces, and is made of sandstone; and the aperture for tobacco had the
appearance of having been drilled out." "We found another skeleton
near that of the above mentioned, which was not quite as large, and must
be that of a woman. There was a neatly carved tombstone near the
head of this skeleton. Close by, the floor was covered with a hard
cement, to all appearances a part of the solid rock, which, after patient
labor and exhaustive work, we succeeded in penetrating, and found it
was but the corner of a box, similarly constructed, in which we found
about three pints of wheat kernels, most of which was dissolved when
brought in contact with the air. A few of the kernels found in the
center of the heap looked bright, and retained their freshness on being
exposed. These were carefully preserved, and last spring planted and
grew nicely. We raised four and a half pounds of heads from these
grains. The wheat is unlike any other raised in this country, and pro-
duces a large yield. It is the club variety; the heads are very long and
hold very large grains." "We find houses in all the mounds, the rooms
of which are as perfect as the day they were built. All the apartments
are nicely plastered, some in white, others in red color. Crockery ware,
cooking utensils, vases—many of a pattern similar to the present age—
are also found. Upon one large stone jug or vase can be traced a perfect delineation of the mountains near here for a distance of twenty
miles. We have found several millstones used for grinding corn, and
plenty of charred corn-cobs, with kernels not unlike what we know as
yellow dent corn. We judge from our observations that those ancient
dwellers of our country followed agriculture for a livelihood, and had
many of the arts and sciences known to us, as we found molds made of
clay for casting different implements, needles made of deer-horns, and
lasts made of stone, and which were in good shape. We also found
many trinkets, such as white stone beads and marbles as good as made

now ; also small squares of polished stones resembling dominoes, but for what use intended we cannot determine."

The above account we see no reason to discredit, and can only wish that the examinations had been more thorough and the account more explicit as to dimensions of rooms and other details. From what is stated, however, we conclude that the authors of these works could not have belonged to the present Indian race, but were undoubtedly of the mound-building people of the Mississippi Valley. It is, at least, a most interesting discovery, and they may belong to a series of structures which shall yet reveal the history of their migrations. That there were two if not three, distinct and widely separated southward movements, in point of time, of the pre-historic race, has already been suggested ; and the Utah mounds may belong to that class which upon further investigation shall furnish the clue to one of the routes pursued, and lead to its demonstration. Should the conjecture as to their authorship be verified, a new chapter of unusual interest in the history of the Mound-builders will be opened for our perusal ; and we may reasonably hope for much valuable information concerning the character and extent of their agriculture, their esthetic taste, and their knowledge of the industrial arts ; and we may find that, in most respects, their social condition was in no wise inferior to that of Mexico and Peru. The wood-carving, plastered and tinted walls, painted vases, and the presence of that most precious of all cereals, wheat, are new and striking evidences of a higher social state than we have hitherto thought possible, whose luxury and refinement were but the presage of a nobler civilization which found its realization and full development in Central and South America, or by some dire calamity was overwhelmed and destroyed.

CHAPTER IX

The works to be described under the head of Historical, or National
Festival Mounds have already been noticed. A representation of one
of this class is given on page 30. (Fig. 9.) It consists of three
embankments placed in a triangular form, enclosing a central mound
which is also enclosed by a circle of small elevation. The ends of
the embankments do not meet, however, but narrow openings are
left at the lines of intersection, and in these openings are found small
truncated mounds. Sometimes, we are told, the group is composed
of two parallel walls, but oftener of three, in triangular position
as just described; while some have been seen which had four embankments arranged in the form of a square; all, however, containing the
central mound with its enclosing circle.

These groups have generally been thought to be defensive works. As
far as known, none have been seen south of Missouri, but it is said they
frequently occur in the States of Iowa, Wisconsin, Indiana, and some in
Illinois. In the latter two States the usual form is square, while in Iowa
and Missouri the triangular arrangement is most frequent. As the walls
are generally of no great height, they are among the first to be leveled
by the plough. But, of whatever form or size, there seems always to
have been observed in their construction a fixed rule in the relative size
of the several parts, whose uniformity invests them with an interest
peculiar to themselves. The group figured on page 30, though found in
Iowa, was selected for description because this form is said to have been
of most frequent occurrence in Missouri.

It will be remembered that the embankments which form the sides of
the triangle were each one hundred and forty-four feet in length, and
respectively three, four and five feet in height, and twelve feet in diameter. The sum of the heights of the embankments is twelve feet, which
is the exact height of the central mound. These multiplied together
equal the length of the embankments—one hundred and forty-four feet.
In all which have been described, the same relation of the several parts
is observed. The embankments are always of equal length, but never

of the same height, while the sum of the heights—whether the group is composed of three or four—always equals the height of the central mound, and the product of both gives the length of the embankments. The tradition concerning them is, that they were erected to perpetuate the union of two or more tribes; the number forming the compact is recorded by the number of embankments, and their relative power by the height of each.   The circle in the center of the enclosure was known as the festival circle, and the small mounds in the angles, or openings, were matrimonial mounds.   To these works the confederated tribes made annual visits, to celebrate the event of their union with singing, dancing and feasting, and a great variety of festive games, which were performed within the enclosure.   The national union thus celebrated was further cemented on these occasions by intermarriages among the members of the different tribes, which took place at the matrimonial mounds. The central mound was known as the union mound, and on festival occasions was occupied conjointly by the chiefs and prophets of each nation, who presided during the celebration.   Concerning the relative age of this class of works nothing is known, and though the tradition above given may be regarded as having no weight or importance, it is quite clear that all conjecture concerning them is equally valueless.

The early writers upon the antiquities of Missouri make frequent mention of the ruins of buildings which were constructed of unhewn stone, and whose walls were said to have been built up with creditable skill and strength, though without durable mortar, if indeed any were used.

Of this kind of structure, the examples are very rare east of the Mississippi.   Whether any are now to be found in any good degree of preservation is quite doubtful.   I will present, therefore, such facts concerning them as can be gleaned from the most trustworthy accounts of early writers.   The first to be noticed are thus described by Mr. Lewis C. Beck, who, after speaking of the pine timber which abounded fifty or sixty years ago along the Gasconade river, and the saw mills erected upon its banks by which the lumber was prepared for the St. Louis market, goes on to state that " near the saw mills, and at a short distance from the road leading from them to St. Louis, are the ruins of an ancient town.   It appears to have been regularly laid out, and the dimensions of the squares, streets, and some of the houses can yet be discovered.   Stone walls are found in different parts of the area, which are frequently covered by large heaps of earth.   Again, a stone work exists, as I am informed by Gen. Ashley, about ten miles below the mills.   It is on the west side of the Gasconade, and is about 25 or 30 feet square; and, although at present in a dilapidated condition, appears to have been built with an uncommon degree of regularity. It is situated on a high bald cliff, which commands a fine and extensive view of the country on all sides. From this stone work is a small foot-path running a devious course down the cliff to the entrance of a cave, in which was found a quantity of ashes. The mouth of the cave commands an easterly view.

"It would be useless at this time to hazard an opinion with regard to the uses of this work, or the beings who erected it. In connection with those of a similar kind which exist on the Mississippi, it forms an interesting subject for speculation. They evidently form a distinct class of ancient works, of which I have, as yet, seen no description."

Another group, described by the same author, was located about two miles southwest of the town of Louisiana. "They are built of stone, with great regularity, and their site is high and commanding, from which I am led to infer that they were intended for places of defence. Works of a similar kind are found on Buffalo creek, and on the Osage river. They certainly form a class of antiquities entirely distinct from the walled towns, fortifications, barrows, or mounds. The regularity of their form and structure favors the conclusion that they were the work of a more civilized race than those who erected the former—a race familiar with the rules of architecture, and perhaps with a perfect system of warfare." The description of those works located near Louisiana is accompanied by a ground-plan or diagram made by the Rev. S. Giddings, a former clergyman of St. Louis, of which Fig. 1 is an exact copy.

DESCRIPTION OF ACCOMPANYING DIAGRAM.

a, b, c, d, outer wall, 18 inches in thickness; length, 56 feet; breadth, 22 feet. The walls are built of rough, unhewn stone, and appear to have been constructed with remarkable regularity.

E is a chamber three feet in width, which was no doubt arched the whole way, as some part of the arch still remains. It is made in the manner represented at 3, and is seldom more than five feet above the surface of the ground: but as it is filled with rubbish it is impossible to say what was its original height.

F is a chamber four feet wide, and in some places the remains of a similar arch still remain.

G is a chamber 12 feet in width, at the extremity of which are the remains of a furnace.

H is a large room with two entrances, I and K. It is covered with a thick growth of trees. The walls are at present from two to five feet in height. One of the trees in the work is two feet in diameter. 2 is a smaller work about 80 rods due east from the former.

A and C are two chambers without any apparent communication with B.

B is a room nearly circular, with an entrance.

In the apartment G, human bones have been found.

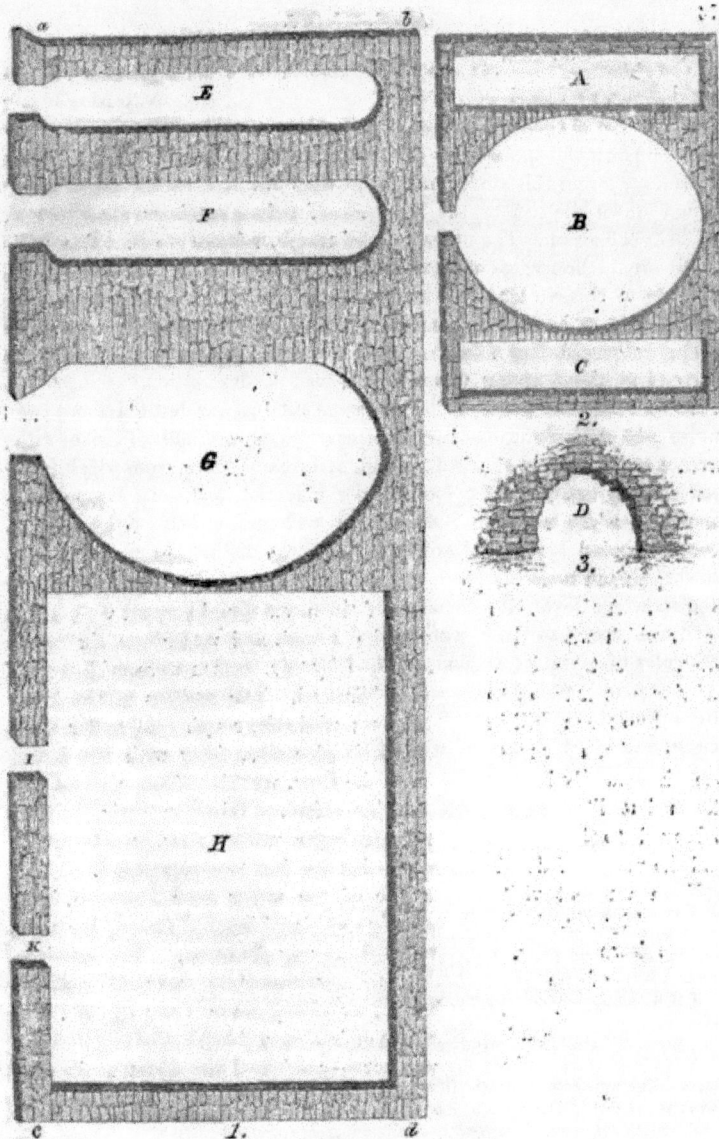

Fig. I.—Ancient Works near Louisiana, Mo.

The stone edifices thus described seem to have been peculiar to Missouri alone, as I find no notices of existing similar works in any other locality, unless those described by Mr. Brown in his *Western Gazetteer* were such. Those were found near the town of Harrisonville, Franklin Co., in the State of Indiana. They were located on the neighboring hills, northeast of the town. The ruins of quite a number were observed, all of which, it is stated, were built of rough, unhewn stone. The walls were levelled nearly to the foundations, and covered with soil, brush and full-grown trees. Mr. Brown informs us that "after clearing away the earth, roots and rubbish from one of them, he found it to have been anciently occupied as a dwelling. It was about twelve feet square. At one end of the building was a regular hearth, on which were yet the ashes and coals of the last fire its owners had ever enjoyed, for around the hearth were the decayed skeletons of eight persons, of different ages, from a small child to the head of a family. Their feet were all pointing towards the hearth, which fact suggests the probability that they were murdered while asleep." The bottom lands in this region are said to have abounded in mounds similar to those described elsewhere, and containing human bones, implements of stone, and a superior article of glazed pottery. A skull taken from one of them was found pierced with a flint arrow which was still sticking in the wound, and was about six inches long. The stone dwellings described by Mr. Brown were evidently of inferior construction to those of Missouri. The authors of the latter showed no mean skill in architecture; while the rough and ruder walls of the Indiana structures, their diminutive size, along with the fact of the whole family lying together on the floor, would indicate a social condition but little removed from barbarism. Whether their builders belonged to the race of the mounds in the valleys near, is not certain, and the means of deciding the question are doubtless destroyed.

Upon a recent visit to the site of the works near Louisiana, Mo., described by Mr. Beck, I found only a confused heap of stones, the walls thrown down and the stones scattered in every direction. The view from the summit of the hill where the building once stood was very extensive and lovely. Mr. Levi Pettibone, now ninety-seven years of age, and Mr. Edwin Draper,—both gentlemen having resided in the neighborhood of the work for nearly half a century—confirmed the account given by Mr. Beck, in every important particular.[1]

---

[1] Mr. Stillman, the obliging and gentlemanly proprietor of the Laclede Hotel at Louisiana, also gave me much valuable information. He stated that formerly there existed

In the February number of the *Western Review* of the present year, appears quite a lengthy article, by Judge E. P. West, containing an account of the examination of several mounds near the Missouri river which contained "buried chambers, or vaults, built of stone, compactly and regularly laid." The stones, which are undressed on the inside, are laid horizontally, and apparently have been selected with great care, the walls presenting, when the earth is removed, a smooth inner face. The chambers were generally of uniform size, being about eight and one half feet square and four feet in height. Each had an opening, or doorway, towards the south, two and a half feet in width. The walls were about eighteen inches in width at the top, and five feet at the base. Some are described as containing "a large quantity of burnt human and animal bones, burnt clay, wood ashes and charred wood, all intermingled and extending entirely over the floor, at one point to the depth of eight inches." Judge West seems to favor the opinion that they were used for dwellings, before the dead were interred in them. This was possibly the case; but the commingled mass of burnt bones, charred wood, and burnt clay to the depth of several inches, would point to funeral rites by cremation. A house eight and a half feet square and four feet high would be a very confined habitation for a family of ordinary size. It seems more in consonance with the facts as stated to suppose them to have been furnaces for consuming the dead by burning. The Judge computes their age to be about two thousand years. Other and similar structures have been described to me, and the localities of their sites named, by respectable persons who claimed to have opened them, of much larger dimensions than any above described, and which are stated to have contained large quantities of human bones and implements of stone. One, I was told, contained a vault at least one hundred and fifty feet in length, fifty feet wide and above twelve feet in height. Another,

upon his land, at a distance of about half a mile from the work described, a stone heap of quite large dimensions, similar in its appearance to those noticed in a previous chapter and conjectured to have been of Indian origin. Having occasion to use the stones for the walls of a cistern, he caused them to be removed. At the bottom of the pile he found a level floor, composed of flat stones of various sizes, but joined together, as he expressed it, as closely and evenly as any mason could do it to-day. From these, and similar facts, I am led to believe that possibly many of those which appear outwardly to be simply piles of stones loosely thrown together, and which are to be counted by thousands upon the hills in various parts of the State, may be the remains of the uncemented walls of ancient habitations. And this conviction receives additional strength from the fact that recent explorations of many earthen mounds have disclosed a vault, walled and arched with stone,—some of large dimensions,—with contents similar to those of Utah.

much smaller, was beautifully arched with stone.   At the time the nar-
rator saw it, it was cleared of the decayed skeletons and was used as a
dairy-house. The two just mentioned were in Missouri, and distant
from each other one hundred and fifty miles.   Again the question recurs
Who built them ; and whence their architectural skill and knowledge?

Says Dr. Foster: "A broad chasm is to be spanned before we can
link the Mound Builders to the North American Indians." There are
some who attempt to do this, but the difficulties which beset the task are
insurmountable to those who have examined, with any degree of thorough-
ness, the evidences of the vastly superior civilization of the people who
erected the stone structures found in Missouri, to that of the North
American Indians, during any known period of their history ; and to
such, the belief that they were the authors of the multitudinous monuments of the Missouri and Mississippi valleys, becomes altogether
improbable.   But if all this is inconclusive of the proposition we maintain, what shall be said of the ancient canals, some of which still remain,
the indubitable evidences of an extended inland communication between
lakes, rivers and bayous, and also of an industry, enterprise and skill
which would be creditabe to the scientific engineers of our own times?
In many of the great achievments of this age of ours we are only recov-
ering the knowledge and wisdom of the long-forgotten past.

When Gov. Clinton, of New York, first proposed the construction of
the Erie Canal, the idea was greeted with scorn and derision ; and as the
work progressed it was characterized as "Clinton's Ditch," the opposers
of the scheme little dreaming that it was to become the great channel for
the commerce of the nation ; connecting, as it does, the great chain of
lakes in the far Northwest with the Atlantic Ocean.   And not until a
thousand freighted boats began to pour the rich treasures of the prairies
into the lap of the East, was the far-seeing wisdom of its projector fully
vindicated.   Then men began to point to it with boasting congratulation,
as an evidence of the rapid and surprising progress which we of the
nineteenth century were achieving.   But alas for human pride ! we are
but slowly learning again what other nations, who lived in the morning
of the historic period, knew, and the world had long ago forgotten.

Again, when the French began the Suez Canal, "all the world won-
dered" at the grandeur of the enterprise.   But they soon found that
they were only clearing out the sands of three or four thousand years'
accumulation from the old pathway of the commerce of the Pharaohs,
who had built the canal when Egypt was the storehouse of the nations.
These came through the canal to her door, in great ships laden with the

riches of the Orient, which they exchanged for corn, and then sailed
back from the Nile, and through the Red Sea to their homes again.
But at length the scepter departed from the throne of the Pharaohs; the
temple colleges, to which the philosophers of Greece resorted for instruction six hundred years before Christ, were closed, and crumbled in decay
—the desert sands swept over their ruins; the canal was filled and
forgotten through all the long dark ages. At length commerce revives,
and men begin to dig canals again, with vain-glorious pride.

It is with nations as with individuals who are taken with some deadly
disease, from which they barely escape with their lives. Though their
strength returns, their memory is utterly oblivious to all they have ever
learned from books, and so they must begin with the alphabet once more.
Nations have their deadly maladies from which few recover, and for those
which do, how long and unpromising is the tutilage of their second childhood. History is repeated here. The pre-historic people of Missouri
were not only great in populous towns, in their agriculture, in their huge
piles of earth and embankments and buildings of stone, but they, too,
were canal-builders. With surprising skill they developed a system of
internal navigation, so connecting the lakes and bayous of the southern
interior of the State, that the products of the soil found a ready outlet
to the great river. The remains of these artificial water-courses have
been frequently alluded to by travelers who have seen them, but never
thoroughly explored. Dr. G. C. Swallow, while at the head of the
Geological Survey, called attention to them, and described one which was
"fifty feet wide and twelve feet deep." For the fullest description of
this class of works, I am indebted to Geo. W. Carleton, Esq., of
Gayoso; who, in response to a note of enquiry,—in addition to many
interesting facts concerning a great number of ancient structures in
Pemiscot County,—kindly furnished the following account, which I give
in his own words :

"Besides our Mounds, we can boast of ancient canals. Col. John H.
Walker informed me that before the earthquakes, these canals—we call
them bayous now—showed very plainly their artificial origin. Since the
country has become settled, the land cleared up, the embankments along
those water courses have been considerably leveled down. One of these
canals is just east of the town of Gayoso. It now connects the flats of
Big Lake with the Mississippi river. Before the bank crumbled off,
taking in Pemiscot bayou, it connected this bayou with the waters of Big
Lake. Another stream, that Col. Walker contended was artificial, is
what we now call Cypress Bend Bayou. He said that it was cut so as to

connect the waters of Cushion Lake with a bayou running into Big Lake. Cushion Lake lies in the northern part of Pemiscot county. The canal was cut from the flats of the lake on the south side, about three miles into Big Lake bayou. By this chain of canals, lakes and bayous, these ancient mound-builders and canal-diggers could have an inland navigation from the Mississippi river at Gayoso, into and through Big Lake bayou and the canal into Cushion Lake, through Cushion Lake and a bayou into Collins Lake or the open bay, thence north through a lake and bayou some eight miles, where another canal tapped this water course and run east into the Mississippi river again, some five miles below the town of New Madrid. Col. Walker, in referring to these water-courses, spoke of them only as canals. They show even now a huge bank of earth, such as would be made by an excavation, on the side opposite to the river, so that in case of overflow the water from the river would not wash the excavated dirt back into the canal." [1]

Although in the foregoing account the present depth and width are not given, from it and from the reports of others, there can be no doubt that the ancient inhabitants had constructed with a skill which would do no discredit to our own engineers, a system of connecting canals which must have been necessitated by an extended internal trade, and which required boats of respectable dimensions. The evidences of work of such magnitude as canals, widen the "broad chasm" which is to be spanned before we can link the Mound-builders to the North American Indians, until it becomes an impassable gulf.

[1] In reply to a subsequent note of inquiry as to the length of this water-course, including canal and bayou, Mr. Carlton estimates it to be about seventy miles.

CHAPTER X.

The number of vessels of pottery which have been taken from the mounds in Missouri is prodigious, and almost endless in variety. In an instance which fell under my own observation, nearly, if not quite, one thousand pieces were obtained from a single burial mound; and these were of various sizes and great diversity of form and workmanship. Some of the most characteristic examples will be presented as we proceed. The skill displayed by the pre-historic Americans in everything they manufactured from common clay is vastly superior to that of the ancient civilizations of Europe, to which, in other respects many striking similarities may be traced.

From the fact that few articles which are the products of human ingenuity and skill are more enduring than earthen-ware, this class of antiquities, to the archæologist, is very interesting and instructive. The skill and taste displayed in its various imitative forms, in outline and decoration, give us an insight into some

Fig. 1.

phases of the domestic life, social condition and æsthetic taste of ancient peoples, which can be derived from no other source. Fragments of pottery, to the archæologist, therefore, are the imperishable leaves of a book, inscribed by the truthful hand of humanity, in legible characters, with the precious records of those feelings and tender sentiments which are recorded nowhere else, and which need no translation. Their value is enhanced so much the more by the fact that we possess specimens of these records from every quarter of the globe, and coeval with the remotest civilizations.

The successful attempts of the ancient Americans to imitate the forms of beasts and birds, which they saw every day around them, evince a contemplation, observation and affectionate communion with nature which fills us with surprise.

The drinking vessel molded into the form of an owl, a representation of which is given in Fig. 1, seems, by its frequent occurrence in the mounds, to have been a favorite model. The most common form is the universal gourd-shaped water jug (Fig. 2). These are of various sizes, the largest being from eight to ten inches high, and the largest diameter not exceeding eight inches. Sometimes the body of the jug is more globular on the top than this figure shows. Fig. 3 presents a form of water jug which, as far as my own observation extends, is much more rare than the preceeding. The engraving was made some years ago ; I have since seen a sufficient number to prove that the reconstruction of the neck is correct. From the greater size of the neck I am led to believe that it was an ordinary drinking-vessel ; while the form represented in Fig. 2 is more properly that of a water-cooler, which, when filled, was hung up until the water was reduced in temperature by its slow evaporation through the pores of the vessel, after the manner of the inhabitants of the American tropics at the present time.

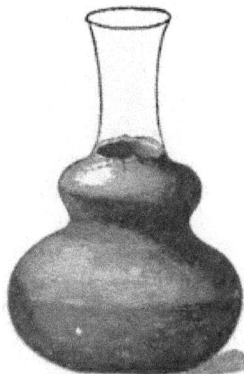

Fig. 2.

In reference to the superiority of the skill displayed by the Mound-builders in the ceramic arts, to the corresponding civilization of ancient Europe, I can not do better than quote the words of Dr. Foster. [1]

"In the plastic arts, the Mound-builders attained a perfection far in advance of any samples which had been found characteristic of the Stone, and even the Bronze Age of Europe. We can readily

Fig. 3.

[1] Pre-historic Races of the United States, p. 236.

conceive that, in the absence of metallic vessels, pottery would be employed as a substitute, and the potter's art would be held in the highest esteem. From making useful forms, it would be natural to advance to the ornamental. Sir John Lubbock remarks that 'few of the British sepulchral urns, belonging to the ante-Roman times, have upon them any curved lines. Representations of animals and plants are almost entirely wanting.' They are even absent from all the articles belonging to the Bronze Age in Switzerland, and I might almost say in Western Europe generally, while ornaments of curved and spiral lines are eminently characteristic of this period. The ornamental ideas of the Stone Age, on the other hand, are confined, so far as we know, to compositions of straight lines, and the idea of a curve scarcely seems to have occurred to them. The most elegant ornaments on their vases are impressions made by the finger-nail, or by a cord wound round the soft clay."

"The commonest forms of the Mound-builders' pottery represent kettles, cups, water-jugs, pipes, vases, etc. Not content with plain surfaces, they frequently ornamented their surfaces with curved lines and fret-work. They even went farther, and moulded images of birds, quadrupeds, and of the human form. The clay, except for their ordinary kettles, where coarse gravel is often intermixed, is finely-tempered, so that it did not warp or crack in baking,—the utensils, when completed, having a yellowish or grayish tint."

In the group of vessels shown in Fig. 4, while the human faces and heads of birds are crudely expressed, we find much to admire in the tasteful forms of the birds themselves. The flow of their outline, so to speak, evinces a degree of refinement of feeling which could only result from a culture of the sense for beauty, which must have required a long time for its realization. It will be noticed that the mouths or openings were, on all, made at the back side of the head. This seems to have been the uniform practice, whether the head of the vessel was that of man, beast or bird. Sometimes the vessels with vertical openings, as of *h* and *l*, are fitted with covers of the same material, with projecting knobs on the top for handling them. Sometimes, again, the smaller jugs, or bottles as they should be called, have nicely-adjusted stoppers, as shown at *i*. These latter bottles are made of much finer material, and while they are generally quite thin, they are so well baked that they seem to be almost as tough and strong as our own ware. On page 23 of the Eighth Annual Report of the Trustees of the Peabody Museum, a representation of two of these stoppers is given; one of which is the same as shown at *i* (Fig. 4). They are described as "two articles

6

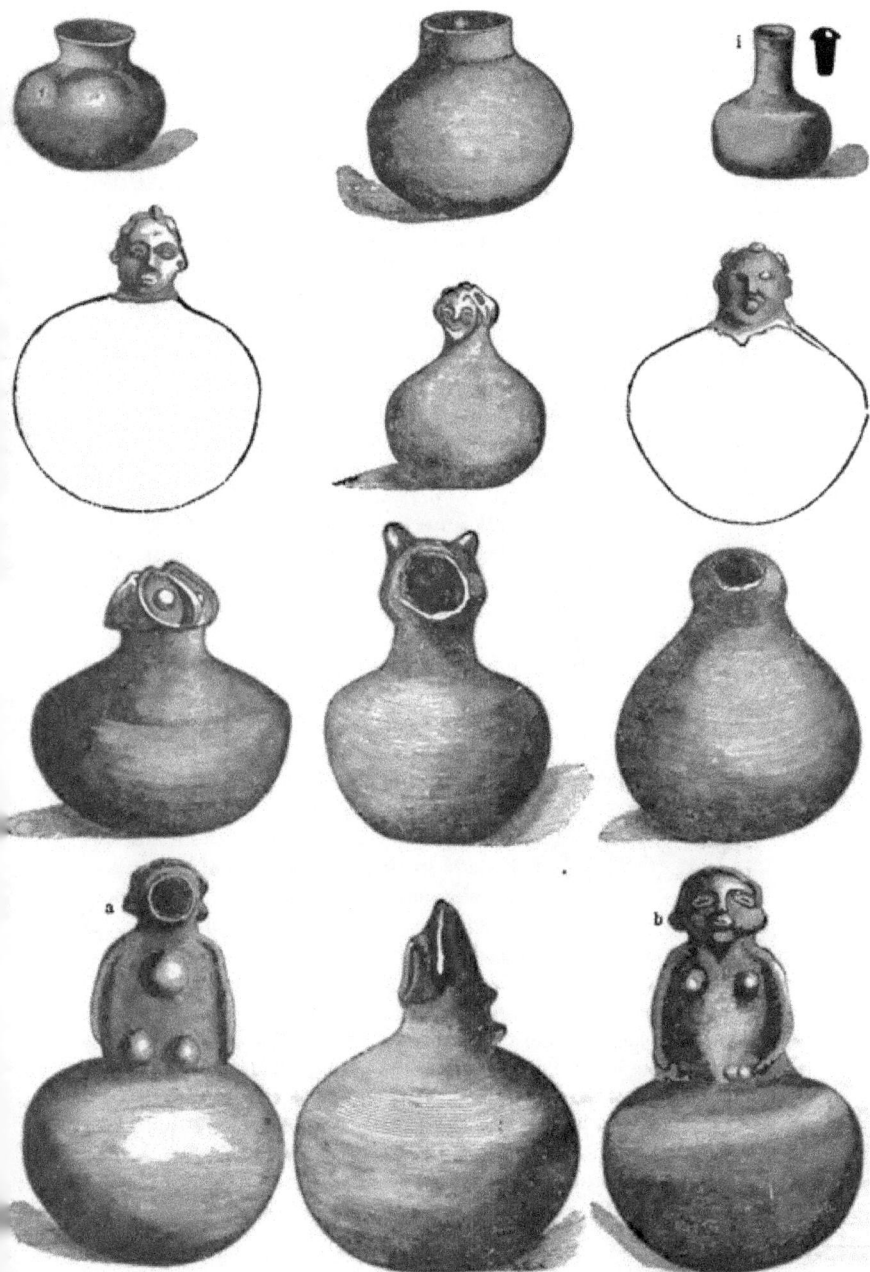

Fig.—4 Varieties of Drinking Vessels from Southeast Missouri. a and b Front and Back View of same Vessel. i Small Bottle and Stopper.

carved from a hard clay slate and carefully smoothed. Their use is problematical, but they so closely resemble lip ornaments as to suggest that they were such." These are now in the "Swallow Collection" of the museum. In its transportation from Missouri to Massachusetts, the report informs us, many of the articles were so broken as to make their reconstruction impossible. When I had the pleasure of examining this collection, some years since, these stoppers were then attached to the bottles with which they were found. The smaller bottle of the two, Professor Swallow informed me, when taken from the mound, contained a red liquid.

Some, of the representations of the human figure are executed with a good degree of fidelity to nature, through all the members; showing that the artist had studied carefully his model, and had evidently labored to tell the truth as he saw it. Some of the human figures have an expression so striking and individual that we can hardly believe that they are not portraits. This becomes more probable when we examine the animal representations, or rather the heads of birds, with which the pottery is very often ornamented; particularly those of the different varieties of ducks, in which we observe in the shape of the head, line of neck, etc., the nicest distinctions in particular varieties, which are expressed with remarkable skill. This will be apparent when we come to the consideration of Food Vessels.

In the annexed group (Fig. 5) are four varieties. In one, the head of the horned owl is skillfully joined to the body of the vessel. Another form of jug, which is of less frequent occurrence than the gourd-shape, is, as shown in the cut, supported by four and sometimes three hollow bulbous legs. The two human figures are coarsely executed, except the heads. They usually represent a hump-backed female figure in a sitting position, and the legs, when they are suggested, bent under the body, with arms resting upon the knees. They are simple water-jugs, having the mouth always in the occipital region of the head. Occasionally one is met with which is grossly indelicate. The vessels representing the human figure vary much in size. Some are so small that their capacity is not greater than two fluid ounces. The larger are from four to ten inches in height

Fig. 5.

and hold from one to four pints. This is, however, a proximate esti-
mate, but can not be far wrong.

Some of the smaller images are, of all that I have seen, altogether the
most artistic and expressive. They have been by some supposed to be
idols, but there is no evidence whatever, that I have seen, which favors this
supposition. They all have an orifice through which the cavities could
be filled, which is constructed precisely like the commonest jugs ; while
their relative position in the mounds, in companionship with other vessels, is conclusive to my own mind that they were used as receptacles of
some precious articles of domestic use ; such as medicines, ointments, and
the like.   And again, there is very
little in all we know concerning this
poeple that would favor the idea
that they had any idols, unless it
may have been symbolic representations of the heavenly bodies,
which we know were the chief
objects of their worship. In addition to all this, they made images
of beasts, as we shall see, which
were unquestionably humorous car-
icatures.

Fig. 6. Two Views.

The most elegant and artistic specimens of pottery which have been
taken from the mounds in Missouri were quite recently discovered.
Some vessels now in the museum of the St. Louis Academy of Science
are very suggestive of the pottery of Ancient Egypt, and indeed, in their
decorative forms, and coloring of black, red and white figures, are not
greatly inferior to Etruscan art.[1]

The material of these articles is much finer than that of the common
ware, which in the larger vessels, having a capacity of several gallons, is
generally mixed with sand, and the medium sizes with pounded shells ;
while the finest seems to be composed of a light-colored, very fine-
grained, yellowish clay—perhaps mixed with gypsum. The different
varieties of ware, the different materials of which they are composed,
and the diversity of tastes displayed in their decoration, would " suggest
a division of labor" among several classes of skilled artisans and artists.

[1] The St. Louis Academy of Science, under the supervision of the Archæological
Section, will soon publish a series of plates of these decorated jugs and vases, drawn
on stone and printed in *fac simile* colors, with descriptive text by F. F. Hilder.

This was probably the case; for, as is well known, however common the articles manufactured may be as to their uses, in everything which comes from the hand of the skillful there is a finish or refinement of treatment which is never seen in the work of the unpracticed hand. The

Fig. 7.

annexed engraving (Fig. 7) represents a jug, about nine inches in height, of a light yellowish color, ornamented around the neck with red and black lines, and around its greatest diameter with curved lines in red, white and black. It is very symmetrical in form, with a bottom sufficiently flat to cause it to stand firmly. I have exhumed one similar in shape and color, but differently ornamented. Around the largest circumference were six red circles; close to these, and on the inside, are white circles. Within these again, is a red circle, and in each of the spaces thus enclosed by the circles, is a white cross with arms of equal

Fig. 8.

length. The stripes are about three-tenths of an inch in width. This combination of color and form has a striking and not unpleasing effect. The knowledge and feeling evinced by the combination and contrast of

angles and circles, in colors, is certainly quite remarkable. The colors
of the stripes were mixed with some sort of article which preserved them,
and gave them a lustrous or varnished appearance, which they still to
some degree retain.

In the next group (Fig. 8) are presented a few of the endless forms
of the more common utensils. They are interesting as showing the
constant and active presence of the inclination to beautify whatever
vessel they manufactured. There are very few that are not ornamented
in some manner. Some have the edges indented or dotted, as with the
point of a stick or the finger-nail, while others have the rim slightly
enlarged and marked with a spiral line, which gives the edge a beaded
appearance. Some of these bowls and pans have a very familiar look as
to their form.

This class of pottery, as well as the ordinary jugs, are usually of dark gray and well
baked, the clay, as before stated, having been
tempered with pounded shells.

In a previous chapter, describing the mode
of burial in one of the mounds near West
Lake, it was stated that with the skeletons
were usually found two or three vessels, one
or two jugs near the head, and a food-vessel

Fig. 9

in the bend of the arms, which were folded across the breast. The
forms of food-vessels here presented are those most frequently found
in that position. In some of them I have observed a very small pot,
not much larger than a hen's egg: in some instances containing a
bone. In others carbonized fruit, resembling wild grapes, has been found;
in others, again, the soft remains of muscle shells, thoroughly decayed.
The jugs and bowls which were interred with the corpse, no doubt, contained food and drink, for the purpose of sustaining the traveler during
the long journey he was supposed to have entered upon. These pots
suggest many interesting reflections concerning their faith and notions
of a future life.

The forms represented in the preceding group are the simplest of all
but not more frequent than those which are much more ornamental.
Vessels in the form of the muscle-shell, and holding fully one pint, are
by no means unfrequent; and again a fish or frog will be used as a
model. The two presented in Fig. 9 are quite common. Sometimes
the legs and feet of the frog are well defined, but folded along the sides
of the body. Usually, when a fish is represented, it is done by simply

moulding the head, tail and fins upon the side of the dish, but occasionally the exact form of the fish is represented, scales and all. In such cases, the orifice is in the side, and furnished with a tube which projects an inch or two, forconvenience in use as a drinking-vessel. In one instance, which came under my notice, the body of a man lying upon the back was represented, with legs and arms rudely made out, and the tube projecting from the stomach.

Fig. 10. Cooking Vessels.

Their imitative faculties, as illustrated in their pottery, were certainly remarkable, and to give an adequate idea of the variety of their work in the subjects which might be chosen for illustration would require more space than is allotted to this essay. We proceed, therefore, to consider their cooking utensils. Some of the more frequent forms are grouped together in Fig. 10.

While these vessels were doubtless for common, every-day use, some of them are really quite artistic and graceful. The three larger ones (a, b, c) are particularly so. The forms and ornamentation of the others seem to be more experimental, and perhaps transitional, as though the maker varied a little from his usual manner just to see how they would look. The one at g, however, is a much bolder innovation, and is finished as there shown, with six hemispheroidal projections. It will be observed that all have two or more handles, by which they were probably suspended over the fire by passing through them green twigs, which they covered with moist clay to prevent them from burning. Examples might be multiplied, ad infinitum almost, of this class of vessels, but the above are sufficient to illustrate the inventive powers of their authors in this direction, as well as their constant striving to gratify their æsthetic feeling in the manufacture of those fragile articles which were designed for the commonest uses.

Fig. 11 represents a pot very similar to a, of the preceding group, but entirely unique in this, that it contained the upper portion of a human skull and one vertebra. It was taken from a mound near New Madrid, by Prof. Swallow, who tells us that the vessel must have been moulded around the skull, as it could not be removed without breaking the pot. It is now in the Peabody Museum. The top of the skull is shown in the engraving. This is certainly a curiosity. Nothing like it has been found in any other burial mound here or anywhere else, as far as known.

Fig. 11.

It may be remembered, however, in this connection, as before remarked, that small pots have frequently been found in the larger pans, and which contained a decayed shell or fragment of bone. These were, very likely, valued relics or charms which were buried with their possessor.

In Herbert Spencer's "Principles of Sociology,"[1] in the chapter upon Idol-Worship and Fetich-Worship, the following interesting statements occur, which seem quite pertinent in this connection:

"Facts, already named, show how sacrifices to the man recently dead

1 Popular Science Monthly for December, 1875, p. 158.

pass into sacrifices to his preserved body. We have seen that to the corpse of a Tahitian chief daily offerings were made on an altar by a priest; and the ancient Central Americans performed kindred rites before bodies dried by artificial heat. That, along with a developed system of embalming, this grew into mummy-worship, Peruvians and Egyptians have furnished proof.

"Here the thing to be observed is that, while believing the ghost of the dead man to have gone away, these peoples had confused notions, either that it was present in the mummy, or that the mummy was itself conscious. Among the Egyptians, this was clearly implied by the practice of sometimes placing their embalmed dead at table. The Peruvians, who by a parallel custom betrayed a like belief, also betrayed it in other ways. By some of them the dried corpse of a parent was carried round the fields that he might see the state of the crops.

"How the ancestor, thus recognized as present, was also recognized as exercising authority, we see in the story given by Santa Cruz. When his second sister refused to marry him, 'Huayna Capac went with presents and offerings to the body of his father, praying him to give her for his wife, but the dead body gave no answer, while fearful signs appeared in the heavens.'

"The primitive idea that any property characterizing an aggregate inheres in all parts of it, implies a corollary from this belief. The soul, present in the body of a dead man preserved entire, is also present in preserved parts of his body. Hence the faith in relics. Ellis tells us that, in the Sandwich Islands, bones of the legs, arms, and sometimes the skulls, of kings and principal chiefs, are carried about by their descendants, under the belief that the spirits exercise guardianship over them. The Crees carry bones and hair of dead persons about for three years. The Caribs, and several Guiana tribes, have their cleaned bones distributed among the relatives after death. The Tasmanians show 'anxiety to possess themselves of a bone from the skull or the arms of their deceased relatives.' The Adamanese 'widows may be seen with the skulls of their deceased partners suspended from their necks.' This belief in the power of relics leads in some cases to direct worship of them. Erskine tells us that the natives of Lifu, Loyalty Islands, who 'invoked the spirits of their departed chiefs,' also 'preserve relics of their dead, such as a finger nail, a tooth, a tuft of hair, and pay divine homage to it.' Of the New Caledonians, Turner says: 'In cases of sickness, and other calamities, they present offerings of food to the skulls of the departed.' Moreover we have the evidence furnished by conversation with the relic. Lander

says : ' In the private fetich hut of the King Adolee at Badagry, the
skull of that monarch's father is *preserved in a clay vessel* placed in the
earth.' He 'gently rebukes it if his success does not happen to answer
his expectations.'

Fig. 12.   Bowls With Ornamental Heads.

" Similarly, Catlin describes the Mandans as placing the skulls of their
dead in a circle.   Each wife knows the skull of her former husband or

child, 'and there seldom passes a day that she does not visit it, with a dish of the best cooked food. There is scarcely an hour in a pleasant day, but more or less of these women may be seen sitting or lying by the skull of their child or husband, talking to it in the most pleasant and endearing language that they can use (as they were wont to do in former days) and seemingly getting an answer back.'

"Thus propitiation of the man just dead leads to propitiation of his preserved body or a preserved part of it; and the ghost is supposed to be present in the part as in the whole."

From the foregoing remarks and array of facts presented by Mr. Spencer, there can be but little doubt that the presence of the skull in the earthen vessel from the New Madrid mound is due to a belief in the presence of the soul in the relics of the departed, and which seems to have been a common belief among many savage and uncivilized nations.

In the next group (Fig. 12) are presented a few of the most common varieties of another and quite distinct class of bowls. They are peculiar in this: the bodies of the vessels are entirely devoid of ornamentation. From the edge of the lip on one side projects a small handle: on the opposite side is moulded the head of some beast or bird, and quite often a human head is represented.

The thing to be specially noticed is the diversity of form in the heads of the ducks. So faithfully are the distinctive features of the different varieties delineated, that those at all familiar with them must believe that the artist, according to the best of his skill, conscientiously copied nature. The beautiful curve of the neck, and its union with the outline of the vessel itself, could not possibly have been accidental.

The best which these ancient workmen could do is so far inferior to the art of our own times, that it is not easy for us to appreciate the difficulties they must have overcome, their many failures, the long time necessary for the acquisition of those habits of observation, and the development of the skill of hand sufficient to enable them to express themselves as creditably as they have done in all their imitative work. In the class of vessels under consideration, examples decorated with the human head and features are by no means rare. If the credit given them for conscientious observation of nature, and skill in expression of what they saw, is not an over-estimate, then we may believe that, in their delineation of the human face, they also copied nature with a sufficient degree of accuracy to warrant us in the idea that in their work we have at least characteristic likenesses of themselves. In the examples presented in Fig. 13, there is wanting that refinement of feeling and realistic portraiture which are displayed in the preceding representations of animal heads; but still sufficient individuality to make them very interesting, and, as before remarked, to impress us with the belief that they too were copied from life.

In the examples thus far given of the pottery of the Missouri Mound-builders, the aim has been to show the leading varieties and mode of decoration. The subject is by no means exhausted; in fact, almost every mound opened discloses some new variety, and I have seen many other specimens of their ware entirely different in form—some of them are beautifully decorated—but which are now scattered among private collections, and therefore not available for illustration here. There is one other curious form of drinking vessel which should be noticed. It has elicited much speculation as to what it was intended to represent.

Fig. 13.　Bowls With Human Heads.

Several of this variety have been found in the Missouri mounds, unmistakably representing the same animal, but no two alike. The general figure of this "what is it" is shown in the engraving. It has four clumsy legs, a thick body, the usual drinking neck projecting from the back, and a swinish head. Sometimes they are made of very fine and finely-tempered yellowish clay,—the larger ones of the usual material of the dark gray ware, with a capacity of from one to two pints. The light-colored and finer ones are decorated with scroll-work made out with red and white lines. Some

Fig. 14.

of the larger ones have human faces moulded upon the sides of the body, midway between the legs. In some instances the head proper has the eyes of a human face and the snout turned up to such an extent as to completely obstruct the front line of vision, which, with its half-human expression, make it very grotesque. If the hog were indigenous to America, it would at once be pronounced a representation of that animal. The nearest approach to it which is native here, is the peccary, or Mexican hog, but that has no tail, while on one example of this figure a tail was well represented; and as it would have been too easily broken in the natural position it was curled up on the hip. Some have pronounced it the hippopotamus. To my own eye it is intensely hoggish. But whatever was intended to be represented by it,—hog or hippopotamus,—it introduces a disturbing factor into the question of chronology which may require some time to adjust; unless we can credit La Vega's statement in his Royal Commentaries of Peru, that the ancient Peruvians who dwelt in the mountains had hogs similar to those which the Spaniards introduced. Again, if the model after which these were moulded was the common hog, which was introduced by the first white settlers in this region, why is it that they took no notice of any other animal or bird which the earliest settlers brought with them, or why do we not find in companionship in the mounds some other human vestigia of European origin? For the present we can only state the facts, with the questions which they suggest, and wait for further developments.

Writers upon American archæology have been able to find no evidence that the Mound-builders knew anything about the use of the potter's wheel; but it is difficult to believe that some of the finest of their work could have been so gracefully and symmetrically moulded by ordinary manipulation, and without some mechanical appliances and adjustments, by which a uniformity of action and pressure would be brought to bear upon the whole mass. Without discussing the question, however, I desire simply to call attention to two discoveries, which at first sight may seem unimportant, but after all may have some value, should they stimulate further and more careful observation in this direction. The first

Fig. 15.

is represented in the engraving, Fig. 15, and was taken from a New Madrid mound by Prof. Swallow. "It is one-half of a rough ball of burnt clay, about 3.5 inches in diameter, and shows the impression of the skin and finger-marks of the hands that moulded it. This mass was perforated through the center, as shown in the figure giving a section of it."[1] It had perhaps been designed to be fashioned into a vessel of some

[1] Eighth Annual Report of the Trustees of the Peabody Museum.

sort, but by some means burnt before the design was carried out. The
perforation would suggest that it had been attached to a stick or spindle
for convenience in handling. The other article is much more suggestive.
It belongs to that class of implements usually denominated spindle-whorls.
They are found scattered over the whole country, at least wherever the
principal works of the Mound-builders are to be seen. This was taken
from a mound about eighteen miles from New Madrid. When I attempted
to wash it, I discovered that it had not been hardened in the fire, but
only sun-dried, as it fell into fragments under
the action of the water. With great care, these
were collected and glued together again. It is
about 2.5 inches in diameter, and three-fourths
of an inch in thickness at the periphery. Both
sides are concave. The most interesting fact about
it is this: It has around the outer edge a rudi-
mentary groove, as represented in the engraving.
Fig. 16. I can only wish the groove had been
deeper. But as it was unburnt, I am led to believe that the article was unfinished; and that
had it been, it would have furnished some evidence that the maker was
not unacquainted with the use of the pulley, or potter's wheel.

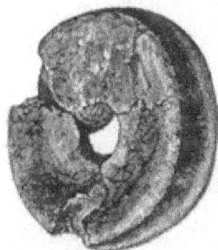

Fig. 16.

The necessity for condensation demands that here our consideration of
this part of our subject should end. The variety and beauty of many of
the objects of their fictile skill are very suggestive, and furnish much
material for extended generalization. But a remark or two must suffice
in this connection. To suppose that all this taste and feeling—this close
observation of nature and fidelity in delineation, displayed in the pottery
of the Mound-builders, found no expression in any other direction, and
was expended upon their domestic utensils alone, is simply incredible.
Very different must have been the homes of a people furnished with
such tasteful articles, from those miserable huts which the nomadic
Indians constructed for their habitations; and it is quite likely that in
their dress as well as their dwellings they evinced the same ideas of taste
and convenience which we perceive in their domestic utensils. In some
of their human effigies we do find the manner of arranging the hair distinctly delineated, and we may yet discover those which shall furnish us
with correct representations of their mode of dress. Indeed I have seen
one vessel with figures of men rudely painted in outline upon its sides,
who were clad in a flowing garment, gathered by a belt around the
waist, and reaching to the knees. In this connection I may mention the

engraved shells which have frequently been found with skeletons, both
in Missouri and Illinois. One of the most interesting is represented
in Fig. 17,[1] which gives also the natural size. When taken from the
mound, the shell was quite soft and brittle, and easily cut with the finger-

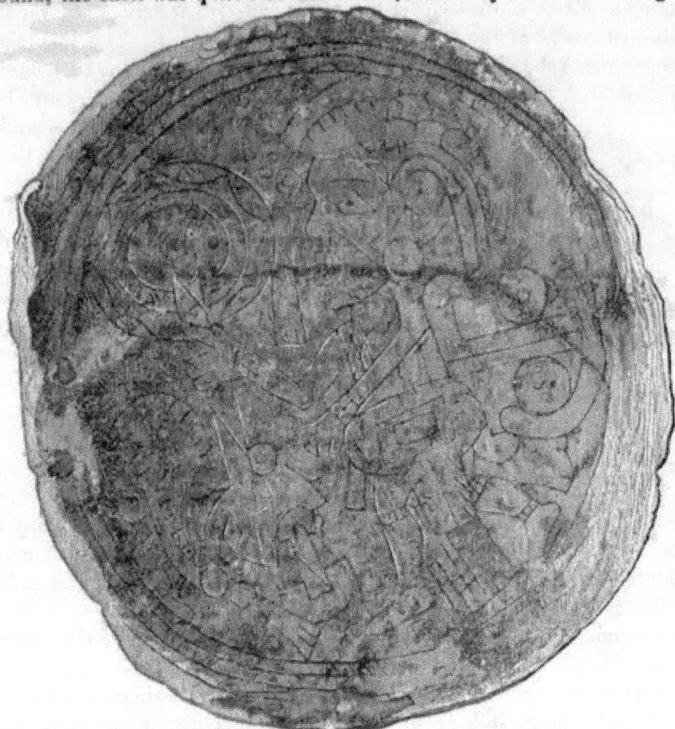

Fig. 17.

nail. The outer edge was much broken or worn away, as shown in the
engraving. The design was enclosed by six circular lines, portions of
which still remain. On one side were two perforations, designed doubtless
for the string by which it was suspended from the neck. All similar
shells that I have seen are so perforated. It seems quite evident from
the picture that it memorializes the victory of the individual represented
as standing over an enemy who lies on his face at his feet. The victor,

[1] For the photograph of which this is an accurate copy, I am indebted to the late
Captain Whitley.

it will be observed, holds in his right hand a weapon or symbol of authority, with which he seems to be pressing the prostrate figure to the earth. Many of the accessories are unintelligible. While the whole work is very crude, and the figures out of all proportion, there is here and there an outline which shows earnest endeavor; as the leg of the standing figure, for example, in which also the action is so well expressed as to suggest that, by an impetuous onset, he has just felled his antagonist to the ground. The artist seems to have had most difficulty with the eye, or rather, has made no attempt at imitating that organ.

There is now in the museum of the St. Louis Academy of Science a similar shell, upon which is portrayed, in a creditable manner, the figure of a spider. I have also been shown another by Dr. Richardson, from a mound in Illinois, almost precisely like it, and differing only in a small symbolic device, which is carved upon the back of each. Engraved shells are generally found upon the breast of the skeleton, or in such a position as shows that they were originally placed there, and also where they were probably worn during life. According to Mr. Pidgeon, the spider emblem is perpetuated in the mounds far to the north. He describes one which he saw in Minnesota, about sixty miles above the junction of the St. Peters river with the Mississippi, which covered nearly an acre of ground. Upon ascending its highest elevation, he tells us, it was very evident that the spider was intended to be represented by it. I bring these facts together for the benefit of future observers, without speculating as to their significance, further than to venture the remark that they point to a great diffusion of one people, or their migration from the north, southwardly along the Mississippi valley.

CHAPTER XI.

CRANIA.—DIFFERENCES BETWEEN THE SKULLS OF THE MOUND-BUILDERS AND THE INDIANS.
—DIFFICULTIES OF THE SUBJECT.—TWO VARIETIES OF CRANIA IN THE SAME MOUNDS.—
PRINCIPLES OF CLASSIFICATION.—INFLUENCE OF LOCAL CUSTOMS.—PERUVIAN SKULLS.—
CHARACTERISTICS OF MISSOURI SPECIMENS, ETC.—THE TOOLS OF ANCIENT AMERICANS.
—PROOFS OF A KNOWLEDGE OF IRON.

To the common observer, the unnumbered stars which shine nightly in the firmament above utter no voice, and give no sign concerning their physical condition, their individual motions, or relative distance from each other. All seemingly sweep on together with undeviating regularity—differing only in the intensity of their light. But when the appliances of modern science are brought to bear upon the facts within our grasp concerning them, and their dim rays are gathered up by the spectroscope, the faint star becomes a fiery orb and the theatre of the conflict of forces of prodigious power. The sun is seen to be a fiery, fluid mass, in whose atmosphere are ceaseless storms of flaming elements and tempestuous cyclones, which burst forth on every side with awful grandeur and inconceivable velocity.

Alike unintelligible to a common observer, in their ethnic relations, would be a collection of skulls brought together from different lands, as throwing any light upon the long history of the different races of mankind. Some would appear shorter, rounder or more irregular than others, but the same general features which characterize them all—with the exceptions named—would be about all that would be specially noticed. But when viewed in the resultant light of all the study which has been bestowed upon them, and the cautious inductions of the wisest ethnologists, they become vocal with revelations of transcendant interest. We are not to suppose, however, that there are no great and decided variations in the crania of a particular race, for these are as widely different as the varying expressions of the human face, and yet all the while presenting certain broad distinctions and characteristics by which the particular race to which they belong may generally be determined.

Says Dr. Foster: " While the individual variations in the crania of a particular race are so great as to present intermediate gradations from one extreme to another, thus forming a connecting link between widely separated races, yet, in a large assemblage of skulls derived from a particular race, there is a general conformation, a predominant type ; which appears

7

to have been constant as far back as human records extend; to have been unaffected by food, climate, or personal pursuits; and which has been regarded among the surest guides in tracing national affinities. Hitherto, our knowledge of the mound-builders' crania has been exceedingly scant—restricted to less than a dozen specimens—which, if authentic, clearly indicate for the most part the Indian type. The results of my observations have led me to infer that the mound-builders' crania were characterized by a general conformation of parts, which clearly separated them from the existing races of man, and particularly from the Indians of North America."

While the number of authentic skulls from the mounds has been greatly multiplied since the above was written, not much has yet been done in the way of classification, measurement and tabulation, so as to be available for serious study. But enough has been already determined to show how premature were the broad generalizations of Dr. Morton—and others who accepted his opinions—deduced from the few examples of the crania of the mounds which he was able to add to his large collection of other types from all parts of the world. While questioning some of his conclusions with which he sums up the results of his long-continued labors, no contrary deductions can detract in the slightest degree from the inestimable value of his labors and splendid contributions to ethnological science. While many, in view of more extended observations and discoveries since his time, will withold their assent to the proposition, "that the American nations, excepting the polar tribes, are of one race and one species, but of two great families, which resemble each other in physical, but differ in intellectual character," all will heartily subscribe to the statement of Dr. Daniel Wilson that, "following in the footsteps of the distinguished Blumenbach, Dr. Morton has the rare merit of having labored with patient zeal and untiring energy, to accumulate and publish to the world the accurately observed data which constitute the only true basis of science. His *Crania Americana* is a noble monument of well-directed industry; and the high estimation in which it is held, as an accurate embodiment of facts, has naturally tended to give additional weight to his deductions."

Nor was this great naturalist less mistaken in his opinion as to the mode of burial practiced by the aborigines of the American continent. He tells us "that from Patagonia to Canada, and from ocean to ocean, and equally in the civilized and uncivilized tribes, a peculiar mode of placing the body in sepulture has been practiced from immemorial time. This peculiarity consists in the sitting posture." That this was not the universal, nor even the most common mode of burial, those who have read the foregoing accounts of explorations in burial mounds in various parts of the continent, have already seen.

He found some difficulty at first in reconciling the peculiarities of the long and flattened Peruvian skulls with the round-headed type of the red Indian, but finally decided that these were only variations of the same type produced by artificial pressure in infancy. But the evidence is abundant and convincing that there was one race in Peru—probably older than the Inca race—with which this peculiarity was not artificial, but congenital, and the skull of the adult retained through life the strangely elongated shape with which it entered the world. Dr. Wilson further remarks in this connection: "The comprehensive generalization of the American cranial type, thus set forth on such high authority, has exercised an important influence on subsequent investigations relative to the aborigines of the New World. It has, indeed, been accepted with such ready faith as a scientific postulate, that Agassiz, Nott, Meigs, and other physiologists and naturalists adopted it without question, and have reasoned from it as one of the few well-determined data of ethnological science. It has no less effectually controlled the deductions of observant travellers."

With such examples before us, a becoming modesty should characterize the conclusions of those laborers in the same great field, who at best may only hope to contribute a page or two to the volume of truth which he has bequeathed to his followers.

The caution with which we should proceed in every step of our investigations becomes all the more imperative on account of the difficulties which meet the observer at the very threshold of his enquiries. One of the difficulties has been already suggested, which is the small number of skulls concerning which there can be no doubt whatever, that they belonged to the race of men who erected the mounds. While it was the custom of the Indian tribes to bury their dead in the mounds which they found ready made, yet their interments may generally be easily distinguished from those of the race of the mounds themselves by the shallowness of the graves, which are usually near the surface. Still, for the want of close observation among cranial collectors, and attention to this fact, much confusion has been the result. Another perplexity is caused by the fact that in the same burial mound are sometimes found—at least in Missouri—two entirely different classes of skulls, with distinctions almost as strongly marked as those which pertain to the Caucasian and Negroid types, whose position in the mound and companionship in the

way of implements and utensils invest neither class with any distinctive claim over the other, as being the individuals for whom the memorial
was erected. But, not to specify further, it may be remarked that so
great are the perplexities caused by these disturbing elements, in the minds
of some, that they have been led to question whether we are justified in
assuming that we have a predominant cranial type of the Mound-building
race, with characteristic conformations so constant as to distinguish them
from all others, wherever found, so that they may be relied on as sure
guides in our investigations.

Assenting, as I do, to the conclusions of such distinguished naturalists
as Wilson and Foster, to the effect that we are justified in assuming that
the crania from localities so far asunder as Illinois, Iowa, Indiana, Ohio
and Missouri, present a "similarity of type in those crania, apart from the
similarity in weapons of warfare, pottery, personal ornaments and earthworks, which would indicate a homogeneous people distributed over a
wide area," yet, to present representative specimens of the skulls which
have been collected from the mounds which are scattered over such an
extended territory, along with the necessary descriptions, measurements
and illustrations as would be requisite for scientific accuracy and induction, would extend our investigations far beyond the limits of the present essay. We must content ourselves, therefore, with such illustrations
and considerations as are more general in their character, but sufficiently
specific and particular, it is hoped, to make them of some scientific value,
at least in clearing the way somewhat for other observers.

For convenience in the study of ethnic relationships, craniologists have
recognized three distinct classes of skulls under which all are grouped.
The principle upon which this classification is made, is based simply upon
the relation of the breadth to the length of the skull. Taking the length
of a skull to be one hundred, when the breadth is less than seventy-
three to one hundred, it is called Dolicocephalic, or long head ; those
whose proportions are from seventy-four or seventy-nine to one hundred
are termed Orthocephalic, or regularly formed ; those skulls whose proportions are from eighty to eighty-nine to one hundred are called Brachycephalic, or short heads. It may be remarked with reference to the
classification of skulls, that some have been found in Europe presenting
such phenomenal characteristics that another class has been proposed,
called Scephocephalic. But, as it is quite likely that the peculiar elonga-
tion of those classed under this head may have been produced by artificial
means, they need not be dwelt upon here. Concerning the skull known
as the "Scioto Mound Skull", which was taken by Squire and Davis from

a mound in the Scioto valley, and figured and described in their great work, Dr. Morton says it is "perhaps the most admirably formed head of the American race hitherto discovered. It possesses the national characteristics in perfection, as seen in the elevated vertex, flattened occiput, great interparietal diameter, ponderous bony structure, salient nose, large jaws and broad face." This skull was regarded by the discoverers as the one of all others concerning which there could be no doubt that it belonged to the race of the mounds; and other eminent writers have accepted the opinion of the finders. Dr. Foster, however, (because of its decided Brachycephalic form doubtless), says that "any comparative anatomist will instantly recognize it as of the Indian type."[1]

As far as my own observation goes, I am persuaded that those ethnologists who have taken one specific form as the type, rejecting all others which do not closely resemble it, do not make sufficient account of the wide extent of territory in which they are found, the length of time which must have passed while the civilization of the race was being developed, nor the influence of local habits and customs in modifying the osteological conditions of the individual members of communities isolated as they must have been for a long series of years; nor of the recognized fact that "individual variations in the crania of a particular race are so great as to present intermediate gradations all the way from one extreme to another, thus forming a connecting link between widely separated races." The burial mounds of Missouri present well-defined Brachycephalic specimens, often flattened in the occipital region, as well as the longer and more symmetrical Orthocephalic type; and sometimes both are observed in one mound. The assumption, therefore, that the one or the other is the exclusive typal form, cannot be maintained; nor on such a narrow basis can these seemingly wide divergencies in the shape of individual skulls be satisfactorily explained. We may safely conclude, therefore, that the idea that one uniform constant type prevailed during the centuries of the occupancy of the Mound-builders of the vast continent of America, through all its fixed communities, is a sweeping assumption which finds no support from the history of other races of men, nor from the facts which the mounds disclose.

The influence of local customs, as exhibited in the different manner of flattening the skull by related tribes of Indians, is a case in point.

[1] It should be remembered that very many other considerations enter into the account in determining the class to which certain skulls belong besides the proportion of breadth to length. This, however, is the first and most important, and the one which I shall chiefly consider

Included under the general name of Flatheads, are at least twenty different tribes. With some, the head of the child is strapped to the cradle-board until its transverse diameter is enormous, when seen in front or from behind, while the longitudinal diameter is only about half as great. In others the skull is shaped by winding a deer-skin cord around the head, beginning just above the ears and winding in such a manner that a uniform pressure is brought to bear upon the skull, forcing it upward until it assumes a tapering form, almost terminating in a point at the vertex. In others again, the pressure is so applied as to press back the frontal bone to such a degree that the forehead is almost entirely obliterated. Concerning the origin of these diverse customs among affiliated tribes we need not stop to enquire. They are sufficient to prove that peculiar practices, affecting the shape of the skull in contrary ways, do originate in communities dwelling near each other, and are persisted in, notwithstanding their constant familiarity with the different customs of their neighbors. It is not surprising then, we repeat, that mounds a thousand miles from each other, or the same mounds even, should disclose cranial forms presenting distinct and contrasting characteristics.

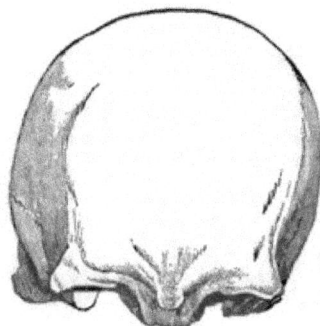

Fig. I.   Front View of Skull from Bayou St. John Mound.

The skull represented in Fig. 1, it will be observed, is very globular in shape, with transverse diameter almost equalling the longitudinal, as will be apparent by comparing the front with the side view which is represented in the next engraving, Fig. 2. From the supercilliary ridges, which are prominent, the line of the forehead ascends almost vertically to a great height, and then sweeps in a well-rounded curve to the apex, from whence it suddenly slopes off in an almost straight line to the occipital protuberance. The squamosal suture is exceptionally straight. The chief point to be noticed in the shape of this skull, is the evidence of artificial flattening seen in the almost straight line from the occipital protuberance to the top of the skull. With few exceptions, all the crania from the Missouri mounds which I have seen are more or less flattened in the occipital regions. Sometimes the pressure seems to have been applied to the right, or to the left of the occipital protuberance, and occasionally directly to the back of the head, and so low down that

the line of the skull from the foramen magnum to the apex of the lamb-
doidal suture is almost vertical. And yet I cannot believe that this
artificial conformation was designed. The absence of any sort of
uniformity in the extent to which it was carried, as well as the indis-
criminate application of the pressure to any part of the occipital regions,
would suggest that it resulted solely from the method of treating the
infant during the first year or two of its existence. The custom of the
North American Indian nomads, of strapping the infant to a board or
basket, for convenience in carrying, and from which it was removed but
seldom until it was at least one year old, need hardly be mentioned.
There is evidence that certain semi-civilized nations so treated their
children as to produce an abnormal shape of their skulls. One reference
must suffice for illustration. Garcel-
lasso de la Vega[1] in speaking of the
manner in which the Peruvian in-
fants were reared, tells us that all
classes, rich and poor, "bred up their
children with the least tenderness
and delicacy that was possible; for
as soon as the infant was born they
washed it in cold water. Their arms
they kept swathed and bound down
for three months, upon supposition
that to loose them sooner would
weaken them; they kept them always
in their cradle, which was a pitiful

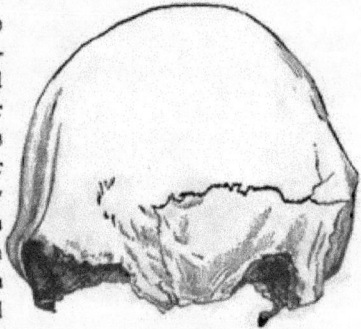

Fig. 2. Side View.

kind of a frame, set on four legs, one of which was shorter than the rest,
for convenience in rocking; the bed was made of a sort of coarse knitting which was something more soft than the bare boards, and with a
string of this knitting they bound up the child on one side and the other
to keep it from falling out. When they gave them suck they never took
them into their lap or arms, for if they had used them in that manner, they believed they would never leave crying, and would always
expect to be in arms, and not lie quiet in their cradles; and, therefore,
the mother would lean over the child, and reach it the breast, which they
did three times a day, that is, morning, noon and night, and unless it
were at these times, they never gave it suck." He tells us in the previous chapter that they were not weaned until they were two years of

[1] Royal Commentaries of Peru, Chap. 12.

age. Some of the Peruvian skulls present a flattened occiput so similar to those of the mounds that it is highly probable this formation was produced by the same means, that is by fastening the infant to the cradle either upon its back, or with the head turned more or less to the one side or the other, in which position it remained until the head became flattened in the region of its contact with the hard bed, thereby receiving a form which it ever afterwards retained.

The skull represented in Fig. 3, when viewed from the front, shows much the same globular form of the brain-case as the preceding ones (Figs. 1 and 2). The vertical view, however, is very different. The flattened portion is more lateral, the pressure having been brought to bear upon the right side of the occiput.

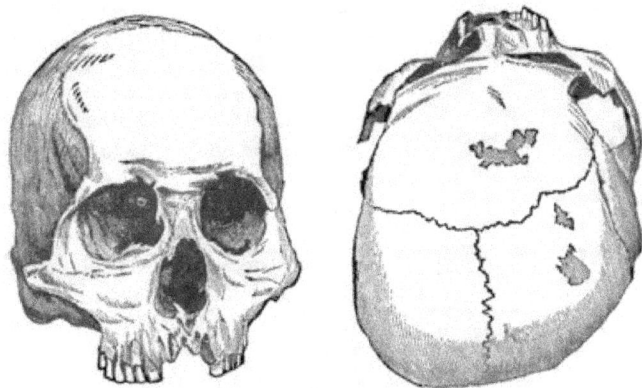

Fig. 3

These decidedly Brachycephalic skulls are very far from conforming to the Mound-builders' type for which Dr. Foster contends. Those represented in his work, taken from mounds in Illinois and Indiana, and undoubtedly authentic, are Orthocephalic or regularly formed. Nor do they present this abnormal deformity of the occiput which characterizes the large majority of those from Missouri. I regret that circumstances forbid the reproduction here of the many cranial forms which are necessary to properly illustrate this part of our subject. But as those figured above, according to the Doctor's views, should be regarded as belonging to the Indian type, I transcribe what he says concerning their peculiarities of form: "The Indian possesses a conformation of skull which clearly separates him from the pre-historic Mound-builder, and such a conformation must give rise to different mental traits. His brain, as compared with the European, according to George Combe, differs widely in the proportions of the different parts. The anterior lobe is small, the middle lobe is large, and the central convolutions on the anterior lobe and upper surface are small. The brain-case is box-like, with the corners rounded off; the occiput extends up vertically; the frontal ridge is prominent; the cerebral vault is pyramidal; the interparietal diameter is great; the supercilliary ridges and zygomatic arches sweep out beyond the general line of the skull; the orbits are quadrangular; the forehead is low; the cheek-bones high; and the jaws prognathous. His character, since first known to the white man, has been signalized by treachery and cruelty." "He was never known volun-tarily to engage in an enterprise requiring methodical labor; he dwells in temporary and movable habitations; he follows the game in their migrations; he imposes the drudgery of life upon his squaw; he takes no heed for the future. To suppose that such a race threw up the strong lines of circumvallation and the symmetrical mounds which crown so many of our river-terraces, is as preposterous, almost, as to suppose that they built the pyramids of Egypt."

In the examples I have given, many of the above traits of the Indian skull are wanting. The anterior lobe is not small; the brain-case is not box-like, nor is the cerebral vault pyramidal; the forehead cannot be said to be low, nor are the orbits quadrangular, or the jaws prognathous. Still, in some other particulars there is a striking conformity to the Indian portraiture. For example, the supercilliary ridges and zygomatic arches in the second example "sweep out beyond the general line of the skull." They are decidedly of the short-head type, and were it not for the de-rangement of the general outline by artificial means in infancy I imagine they would correspond in a striking manner to the Scioto Mound skull, which Foster believes to belong to the red race. The occipital and lat-eral depression shown in the vertical view, Fig. 3, is by no means confined to the skulls of the Missouri mounds, but is found in Peru. If the reader will consult Morton's *Crania Americana*, Plates B and C, he will find skulls with the identical characteristics of the one at Fig. 3. They occur in the mounds of the upper Mississippi region, and in Ten-nessee. In *Harper's Magazine* of December, 1876, is a valuable arch-æological article by Dr. Jones, in which I find the engraving of a skull whose resemblance to Fig. 3 is so striking that I reproduce them both side by side. The thing to be noticed is the general outline in which the similar depression is shown. In the Tennessee skull—assuming that

the same point of view is taken in both—the zygomatic arches are
scarcely seen, while in that from Missouri they bulge out far beyond
the general outline.

While, as before remarked, the majority of the skulls found in the
Missouri mounds possess the characteristics shown in the examples here
given ; some which occur more rarely are so strikingly different that they
can not by any reasonable theory be classed with them.   While exploring
a mound in southeast Missouri, before referred to as having been the
burial-place of many hundreds, two skeletons were found lying beside
each other, so decayed that the bones could scarcely be handled at all
without crumbling to pieces.   The skulls were entire when passed up to
me from the excavation.   They were so peculiar that I was filled with

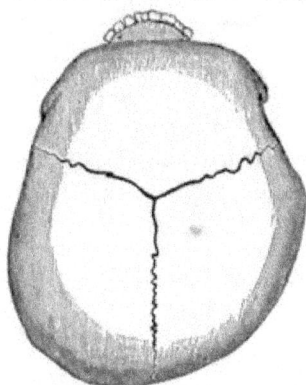

Skull from Mound in Tennessee.   Fig. 4.   Skull from Mound in Southeast Missouri.

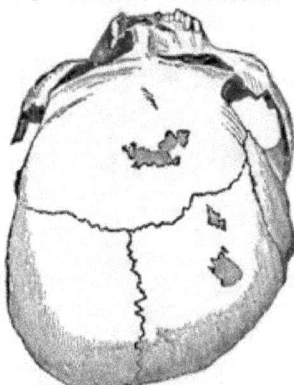

astonishment the moment I saw them.   One crumbled to dust in a few
moments after its exposure to the air, and fell from my hands, along with
the earth with which it was filled, like all others, which are—as well as
bowls and small-necked water-jugs—always densely packed with the loam
which covers them.   I proceeded to a more careful examination, as I
suspected an intrusive burial.   With much painstaking I was able to preserve the upper portion of the second skull, which was a duplicate of the
one destroyed.   The outline of this fragment is well represented in the
engraving.   Both skeletons were lying upon the back, with the head
toward the center of the mound, with the usual drinking vessels close to
the head, and a food-vessel in the angle of the folded arm upon the
breast.   It will be seen at a glance that the forehead is annihilated ; the
frontal sinus is quite prominent, which, along with the almost horizontal

line of the frontal bone, makes this part of the skull resemble that of a beast more than a human head; and yet I am quite sure that its form was perfectly natural, for I could detect no indication of an artificial depression in any part of it. The frontal bone was curved backward, on each side of the occiput, and from the foramen magnum, or from the bottom of the brain-case to the apex, was one graceful curve. It might be suggested —as has been done in the case of the Neanderthal skull—that these were the skulls of idiots. But whoever they were, they were buried with tender care, and in the belief that they would enjoy another life beyond the grave in companionship with the many hundreds of others who were provided with the necessary food and drink to sustain them during their long journey. As so large

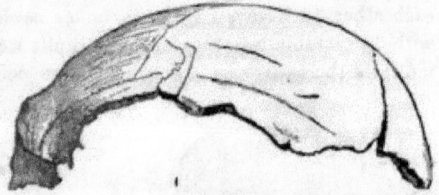

Fig. 5.

a portion of the skull is wanting, it is perhaps useless to generalize upon so small a fragment.[1] Still I can but record my own strong conviction that we have here no idiotic anomaly, but characteristic examples of a race of men entirely distinct from those who piled up the mounds in southeast Missouri. Much evidence is gathered from widely separated localities upon the American continent, which suggests more than the probability that it was once inhabited by a race of men whose origin must ever be hid in the night of oblivion, and the date of whose occupancy may not be far from that of the Paleolithic races of Europe.

The same configuration has been found in the bone caves of Brazil, and in companionship with extinct animals. Dr. Lund thinks they were contemporaneous. In some which he describes, the peculiarities which characterize them are "in excessive degree, even to the entire disappearance of the forehead." The same form appears in the sculptures on the most ancient monuments of Mexico, as also in the bas-reliefs of Uxmal and Copan, in Central America. "Humboldt and Bonpland," says Foster, "were the first to draw attention to this remarkable configuration of skull. The former, as far back as 1808, thus stated: "This extraordinary flatness is found among nations to whom the means of producing artificial deformity are totally unknown, as is proved by the crania of Mex-

[1] For the distinguishing traits of idiotic skulls, consult Humphrey's Treatise on the Human Skeleton, p. 233.

ican Indians, Peruvians, and Atures." Pentland, Cuvier, Gall and
Tiedeman believed this strange cranial form to be congenital. Rogers
and Tschudi both were convinced of the former existence of an Autochthonous race in Peru with this peculiarity of skull, and "state that it is
seen in the fœtus of Peruvian mummies." Dr. Lapham has observed
what seems to be the same type, in Wisconsin. In a private note to Dr.
Foster he says, concerning two skulls found at Wauwatasa: "The peculiar characteristics indicating a low grade of humanity, common to
both, are a low forehead, prominent superciliary ridges, the zygomatic
arches swelling out beyond the walls of the skull, and especially the
prominence of the occipital ridge. The anterior portion of these skulls,
besides being low, are much narrowed, giving the outline, as seen from
above, of an ovate form. It seems quite probable that men with skulls
of this low grade were the most ancient upon this continent, that they'
were the first to heap up those curiously-shaped mounds of earth which
now so much puzzle the antiquary ; that they were gradually superceded
and crowded out by a superior race, who, adopting many of their customs, continued to build mounds and bury their dead in mounds already
built. Hence we find Mound-builders' skulls with this ancient form,
associated with others of more modern type. The discovery of these
skulls, with characteristics so much like those of the most ancient of the
pre-historic types of Europe, would seem to indicate that if America
was peopled by emigration from the Old World, that event must have
taken place at a very early time—far back of any of which we have any
record." The occurrence of skulls with this unique and congenital configuration, in both continents of America, from Wisconsin to Peru,
and many of them associated with those ancient structures whose authors
are unknown to history or tradition, are facts not to be overlooked or
lightly considered in tracing the ethnic distinctions of the pre-historic
inhabitants of the, so-called, New World. They are certainly very
suggestive, and invite the serious study of future observers.

There are certain facts which have been noted from time to time,
which fit into none of the popular theories concerning the state of the
arts of the Mound-builders. It has been stated, and often repeated, that
they had no knowledge of smelting or casting metals, yet the recent
discoveries in Wisconsin of implements of copper cast in molds—as well
as the molds themselves, of various patterns, and wrought with much skill
—prove that the age of metallurgical arts had dawned in that region at
least.

And again: what shall be said concerning the traces of iron implements which have been discovered from time to time in the mounds, but more frequently at great depths below the surface of the soil. Though accounts of such discoveries are generally from reliable sources, they have latterly received no attention, and always have been considered as so much perilous ware which no one cared to handle. The peculiar ovate form of skull with the retreating forehead, as has just been shown, points to the presence, in remote times, of a race of men entirely different from that to which the authors of many of the earthworks of the Mississippi valley belonged. This form has been traced to Mexico and Peru. When the Spanish conquerors pillaged those countries and laid waste their beautiful cities, they observed vast structures and ancient temples built of hewn stone, with consummate skill. When they questioned the Aztec and Inca races concerning their origin they could give no answer but this: they were here when our fathers came; they belonged to a people of whose history we know nothing. The Incas copied these ancient models in the great structures which they erected. But with what tools did they perform such wonders—were they of copper only? So we are told; or copper alloyed with tin. It is said they had some secret method of making it hard as iron, but none of the copper tools which have been found confirm the statement. Mountains of stone were wrought into dwellings and temples of the gods; huge walls were cut from the solid rocks; the mountains themselves divided into galleries and fortifications rising one above the other, connected here and there by artificial breastworks, but generally cut out of the strata of the mountain and left standing, one solid mass of stone. Common dwellings built of enormous slabs of stone seven feet wide and twelve feet long are met with. Porphyry, basalt and marble yielded alike to their magic touch, like clay in the hands of the potter. Vitreous obsidian was utilized by the excellence of their tools and the delicacy of their manipulation. Plates and cylinders of exquisite thinness they made of this fragile substance for ornaments for their women.

The dexterity of these ancient lapidaries in cutting the hardest stone is amazing. And it is difficult to conceive how, without cutting implements equal, at least, to our own in hardness, such delicate and such stupendous works could have been executed. And to the question whether they possessed a knowledge of working iron, the wise man will hesitate long before he answers in the negative. It should be remembered, too, how quickly—unless under most favoring conditions—iron corrodes to dust and leaves scarcely a trace behind. The piles of the Swiss lake-dwellings, the cedar posts of the mounds, may endure for ages, while iron—so hard, and more precious than gold in the advancement of the world's civilization,—speedily melts away before the gentle dews and air of heaven.

The idea that there once existed on this continent a race anterior to, and entirely distinct from, that which immediately preceded the red men, is no new and fanciful conjecture, but one which was held by the earliest and most cautious observers of the antiquities of America; and we may yet be forced to adopt their conclusions, not only upon this point, but also their opinions as to the state of the arts in those remote times.

According to Morse, the geographer: "In digging a well in Cincinnati, the stump of a tree was found in a sound state, ninety feet below the surface; and in digging another well, at the same place, another stump was found at ninety-four feet below the surface, which had evident marks of the axe; and on its top there appeared as if some iron tool had been consumed by rust."

Says Priest: " We have examined the blade of a sword found in Philadelphia, now at Peale's Museum, in New York, which was taken out of the ground something more than sixty feet below the surface. The blade is about twenty inches in length, is sharp on one edge, with a thick back, a little turned up at the point, with a shank drawn out three or four inches long, which was doubtless inserted in the handle, and clinched at the end."

"Twelve miles west of Chillicothe, on Paint Creek, are found the remains of a furnace, ten or twelve feet square, formed of rough stones, surrounded by cinders, among trees of full size. There are, at this place, seven wells situated within the compass of an acre of land, regularly walled up with *hewn* stone, but are now nearly filled up with the accumulating earth of ages. Eight miles further up the creek, a small bar of gold was taken out of a mound, which sold in Chillicothe for twelve dollars. A piece of cast iron, we are further told, was taken from a circular embankment in Circleville."

From the distinguished antiquary, Mr. Atwater, who was present when a large mound near Circleville was removed, we learn that in addition to the skeletons it contained, along with stone implements, was found " the handle, either of a small sword or large knife, made of elk's horn; around the end, where the blade had been inserted, was a ferrule of silver, which, though black, was not much injured by time; though the handle showed the hole where the blade had been inserted, yet no iron was found, but an oxide, or rust, remained, of similar shape and size." With another skeleton, in the same mound, was found a large plate of mica, three feet in length by one and one-half in width, and one inch and a half in thickness. On this was a plate of iron thoroughly oxydized, which crumbled to dust when disturbed by the spade, but resembled a plate of cast iron. This was doubtless a mirror. Both bodies had been burned, and mingled with the bones and implements were quantities of charcoal and ashes. The same author thinks that some of the supposed iron knives which have been discovered in the mounds may have been steel instead. The "huge iron weapon" found in the hand of the skeleton in the Utah mound before described, which crumbled to dust on exposure to the air, will be remembered. But here I must desist from further consideration of the question—for the present at least—as to the extent of the knowledge and uses of iron among the ancient Americans, as I am not aware of any relics of this metal having been found among the antiquities of Missouri, save those made of native ore. But, as similar notices of its occurrence in the mounds and on ancient levels, far below the present surface of the alluvial plains, are abundant in all the current antiquarian literature of the last half century, I felt that the subject was too important to be passed over in silence, especially as I had reason to suspect that those remote dwellers upon this continent, whose peculiar form of skull has been noticed by Humboldt, Foster, Lapham and many others, and lastly by myself in Missouri, were not unacquainted with the uses of iron and other metallurgical arts. That these were the opinions of that distinguished scholar and statesman, William Wirt, the following quotation from his writings will show. After speaking of the various relics of vanished races, among which he mentions "iron and copper, buried in a soil which must have been undisturbed for ages," he proceeds to say: "The mighty remains of the past, to which we have alluded, indicate the existence of three distinct races of men, previous to the arrival of the existing white settlers. The monuments of the first or primitive race, are regular stone walls, wells stoned up, brick hearths found in digging the Louisville canal, nineteen feet below the surface, with the coals of the last domestic fire upon them, medals of copper and silver, swords and other implements of iron. Mr. Flint assures us that he has seen these strange ancient swords. He has also examined a small iron shoe, like a horse-shoe, encrusted with the rust of ages, and found far below the soil, and the copper axe, weighing about two pounds, singularly tempered and of peculiar construction." The second race, he thinks, were the authors of the mounds, who, in time, were succeeded by the Indians.

A few weeks since I received, in a private letter from Prof. Tice, the distinguished metèorologist, an interesting account of the discovery, in one of the interior counties in Illinois, of the corroded remains of some sort of cutting implement of iron or steel. As I have not his communication at hand at this moment, I cannot give the details; but as I recall the statement, it was found several feet below the surface, in a gravelly river bank which had been washed away by the floods and thus exposed, and under such circumstances as to convince intelligent observers who saw it, and the bed from whence it was taken, that it was of great antiquity. What shall we say to these numberless and constantly recurring notices of the discovery of traces of iron? The journey of De Soto across the continent has been made to do good service in explaining the presence in the mounds of metal implements, as well as the immense defensive structures in some of the Southern States, which were thought to be beyond the skill of the ancient inhabitants. A topographical representation of all of the supposed routes of his journeyings would resemble a western railway chart. Had De Soto lived till now, and traveled incessantly, like the Wandering Jew, he could not have accomplished all that has been placed to his credit. Again, the bold Norsemen, under Eric the Red, and other adventurers on the ocean, whose ships, by adverse winds or favoring gales, were driven to these far off shores,—colonies of Welsh, Malayans—and the lost ten tribes of Israel—all have been marshaled by different authors in the interest of their particular theories, and made to do duty in explaining the inexplicable problems of our antiquities. In regard to the question thus touched upon, as well as many others equally perplexing, is it not better to sift and garner the grains of truth we have, and with childlike receptivity wait for greater light?

CHAPTER XII.

A proper completion of our investigations demands a brief notice of
the current opinions which relate to the origin, migrations and the ultimate fate of the race whose relics and monuments have been considered
in the preceding pages. By whatever theory we may be pleased to
adopt as to the manner in which was first peopled, we are carried back
irresistably to times so remote that we rise from our study of this subject with the conviction that the origin of the first inhabitants of
this continent must ever remain hidden in the darkness of oblivion.

None of the many theories, some of which seemed quite probable at
first view, have withstood the test of later investigations. One nation
after another—European or Asiatic—has been put forward, as entitled to
the honor of having been first in the field with its peopling or civilizing
colonies; prior to whose coming, it was assumed, this continent must
have been a desolate waste, without inhabitants, or, in the latter case, at
best, the home of wild and barbarous tribes. Another theory, which is
maintained by a few distinguished writers, is based upon the hypothesis
of spontaneous generation; the natural sequence of which is that the aboriginal inhabitants of America were Autochthons;[1] or in other words, that
man—in common with the plants and lower orders of animals—made his
appearance on the earth spontaneously, when, in the fullness of time, it
had reached that condition which presented all those favoring and con-
current circumstances which made his appearance a natural necessity.

The spontaneous generation hypothesis is still so far from being verified,
that the question of an autochthonous population need not be discussed;

---

[1] Humboldt suggestively asks, "Did the nations of the Mexican race, in their migrations to the south, send colonies towards the east, or do the monuments of the United
States pertain to the Autochthone nations? Perhaps we must admit in North America,
as in the ancient world. the simultaneous existence of several centers of civilization, of
which the mutual relations are not known in history." Personal Narrative, Vol. VI., p.
322.

8

inasmuch, also, as what we have to consider farther, relates to the ancient
people of Missouri; who, whatever may have been their origin, were so
far removed in time from the parent stock, and changed in their physical
and social condition by their evident subsequent commingling with the
Indian tribes, that they furnish us with few, if any, facts which can be relied on as sure guides in conducting us to the origin of their national life.

We must take them, therefore, as we find them, and in the light of such
facts as we have been able to gather, and, applying also the mysterious,
yet well-established law which seems ever to have controlled the migratory movements of the various nations of both hemispheres, deduce such
conclusions as we may be justified in doing concerning their own migrations and their ultimate fate.

The student of ancient history will observe that the migrations of rude
and semi-civilized nations have generally, if not always, been from north to
south. The exceptions to this, which are exceedingly rare in proportion
to the vast number of known movements of tribes and peoples in this
direction, it is believed may readily be accounted for by some local and
temporary cause—as stress of war, for example—which turned them for
a time from their normal course. The constancy of the operation of this
law—the causes of which are yet the subject of much learned speculation—I shall assume without stopping to illustrate it by quoting
the numerous examples with which the pages of history abound, further
than to give the opinion of one distinguished naturalist in its support.
Says Von Hellwald,[1] "If we seek, however, to establish for historical
events a basis in geographical relations—that is to say, if we carefully
compare them together, analyzing the former and investigating their possible causes, studying the latter and deducing as far as possible the
resulting consequences—we shall find that certain generally valid laws,
which resolve in the simplest manner many an unexplained riddle, are
evolved from such a study through the remarkable correspondence of
facts. Thus, in reference to the migrations of mankind, it seems to
result from the geographical structure of the continent that, as by virtue of
an *historical law* we are not to look for men of comprehensive and
deeply penetrating intellect in Lapland or Malta, in Bosnia or Asturias;
so, conformably with a strict *geographical law, the direction of the migratory stream will be found always to lie in the axis of the greatest longitudi-*

_____
[1] The American Migration, by Frederic Von Hellwald—an admirable essay. Some of
his facts and dates I have adopted.
Smithsonian Report, 1866.

nal extension of the continent. In fact, no example from history informs us that the Tchapogires, Tunguses, Jakoots or people from the banks of the Amour, have ever descended into the Deccan or Malacca; that the Ethiopians have ever migrated into Sennegambia, or the Finns into Greece. As a new proof how much nations and men depend on geographical circumstances, and even when they believe themselves guided by their own will, merely obey a great natural law, the fact is of much significance that the American tribes form no exception to this general rule; for here, also, the procession of the migratory races is in the longer axis of the continent, namely, from north to south.

"That America, as well as Europe and Asia, was already inhabited before this great migration, and in many parts possessed of an ancient civilization, admits of no doubt. Occasional traditions of those early periods of culture have penetrated to us, and I cannot forbear soliciting the attention of the learned world to this legendary cycle of America, which is certainly worthy in many respects of a critical scrutiny; for to judge from so much as is yet known, the inquiry cannot but yield interesting and valuable disclosures respecting the cosmogonic views of the American aborigines and the general tendency of their ideas; perhaps endow even the historian here and there with a fact of value. But to determine, from our present knowledge of the mystical traditions of these races, which of the tribes in America may have been the oldest, seems to me as impossible as superfluous.

"Upon this soil multitudes of nations have moved and have sunk into the night of oblivion, without leaving a trace of their existence; without a memorial, through which we might have at least learned their names. Those nations only, which by tradition, written records, monuments, or whatever other means, first guaranteed the remembrance of their own existence, belong to the domain of history; and history which, to be true, accepts nothing but what is actually known, points to those as the primitive races which first transmitted a knowledge of themselves; time begins for us when the chronology of such nations takes its rise. But all these so-called aborigines might be only the remainder of previously-existing races, of whom, again, we know not whether they were indeed the first occupants of the land. In truth we meet in America, at more than one point, with traces of a rich civilization, proceeding demonstratively from much earlier epochs than the tribal migration itself; as, for example, in upper Peru, the gorgeous structures of the Aymaras, near Tiahuanco, on the beautiful shores of Lake Titicaca; the mysterious monuments of Central America, between Chiapas and Yucatan, of which

the buildings of Palenque constitute the most celebrated representative ;
the earth and stone works of a people distinct from the above, on the
banks of the Mississippi and the Ohio."

In speaking of the migratory movements of the American tribes, it
must be remembered that several distinct expeditions of the same people,
at times more or less remote from each other, are often spoken of as one
migration ; for example, the race which bore the name of Nahuatlacas,
was composed of seven tribes ; namely, the Xochimilcos, Chalcas, Te-
panicas, Tlahuicas, Colhuas, Tlaxcaltecas and Aztecs. All these tribes
spoke the same language, and, issuing from the same region far to the
north, appeared in Mexico at successive periods, following each other in
the order named. The Aztecs, renowned in the history of the Conquest,
were the last to arrive. Some time prior to the commencement of the
Christian era—many think not less than a thousand years must be
assumed—the mysterious Nahoas, or Nahuas, appear in Mexico. Concerning their origin little is known, and none have been able to penetrate
the clouds of obscurity which envelop their history. This much, however,
is established, namely, that all the Toltec and the later Aztec, or more
properly Nahuatl tribes, were only branches of the great Nahua family,
and all spoke dialects of this ancestral race. This is a most important
and significant fact, as affinities of language are considered among the
most certain guides in ethnological investigations. But little more is
known concerning the original Nahuas than to suggest the probability
that they were the authors of some of the stone structures in Northern
Mexico, and the builders of a few, and those the most ancient, mounds
of the Mississippi Valley. With the advent of the Toltec domination in
the country previously occupied by their Nahuatl[1] ancestors, the thick
darkness begins to be dissipated, and the dawn of ancient American
history is ushered in.

The learned and able interpreter of the monuments and hieroglyphic
annals of ancient Mexico, the Abbe Brasseur de Bourbourg, regards 955
years before Christ as the earliest reliable date which can be established
in the Nahuatl language. Although the Toltec tribes did not make their
appearance on the scene simultaneously, but at different times, and possibly by different routes, as was the case with the Aztecs who succeeded
them, their active occupancy began in the seventh century of our era, or,
to speak more accurately, in the year 648. Clavigero, however, who is
alone in his opinion among early writers, fixes the date at 596. This

[1] Pronounced "Know-all"; and according to de Bourbourg, it has the same meaning.

people, after the lapse of about four hundred years, having been almost destroyed by famine, pestilence and civil wars, were succeeded by a more barbarous, though neighboring tribe, known as Chichimecs, who also have been supposed by some to have belonged to the same Nahua family, but whose peculiar language is now considered as convincing proof that they were from a separate and distinct stock, although they had been more or less influenced by association with their Toltec **neighbors, and** had adopted some 'of their arts and customs. **Of course it should be** remembered that the large territory of Mexico was occupied conjointly by many other pre-Toltec tribes besides the Nahuas, but whose languages were so radically different, so entirely wanting in linguistic affinities with the Nahuatl tongues,—among which may be mentioned the Almecs and the Otomi, whose speech was monosyllabic,—that they must be regarded as more ancient than the Nahuas even. But the reign of the Chichimecs was short. A tribe of immigrants, known as the Acolhuas, took up their residence with the Chichimecs, and the union resulted in the kingdom of Acolhuan. **This kingdom was scarcely established when the great and** last migrations we have to notice took place. The seven Nahuatlacas tribes, as before noticed, arrived upon the scene, the Aztecs bringing up the rear, after a longer interval than the others. This celebrated people, who, in the year 1090, had left their home in the mysterious Aztlan, after various wanderings and delays in their southward journey, finally reached the table-lands of Mexico somewhere between the years 1186 and 1194, and took possession of the cities which the Chichimec in turn abandoned, following in the path of the Toltecs, who had fled from these same seats less than two hundred years before. Adopting and improving upon the civilization of their predecessors, the Aztecs founded that kingdom whose magnificence and power filled the Conquerors with wonder. They displayed a bravery and heroic devotion in the defense of their rulers and their native land which awakens our liveliest sympathies, and the admiration of the civilized world; and their final and pitiless destruction has left a dark stain upon the character of their destroyers, which no excuses in the interests of religious zeal can diminish, nor the glory of their daring deeds efface!

In the preceding and incomplete outline sketch of the leading branches of the Nahua family, with some account of their migrations, I have called attention to those facts only which seemed necessary to a more explicit statement of what has been incidentally assumed throughout these investigations.

From my point of view then, no theory is admissible which does not contemplate the migrations of the various tribes which appeared at different times upon the table lands of Mexico, during a period of two thousand years or more, as the movements of the different branches of the one Nahua race, whose ancient seats must be sought for in the great alluvial plains of the Mississippi valley. Their precise location may never be discovered; it is, however, quite probable that the unknown Aztlan, the Huehuetlapalan of the Aztecs (who, as has been shown, were the last to leave their primitive home), may yet be identified. At the commencement of my study of the antiquities of America, I accepted without question the views of distinguished early writers upon this subject, which I have since found no reason to reject during all my subsequent inquiries. And had I at any time been disposed to embrace opposite conclusions, I should have felt great diffidence in suggesting them, which to me would savor of presumption, thus to place myself in opposition to the mature convictions of the great men who have devoted years of patient labor in this direction, of whose names I need mention but one. Among the learned in all lands, the opinions of Humboldt, upon any subject which engaged the attention of his powerful intellect, command the most respectful consideration. The rare opportunities which he enjoyed, during his extended travels and prolonged stay on this continent, at a period, too, when many of the antiquities were in a better state of preservation and therefore much more intelligible and instructive than now, give great weight to his conclusions concerning the ancient races of America. In speaking of the races under consideration, he says: "The very civilized nations of New Spain, the Toltecs, the Chichimecs, and the Aztecs, pretended to have issued successively, from the sixth to the twelfth century, from three neighboring countries situated toward the north, and called Huehuetlapalan or Tlepallan, Amaquemacan, and Aztlan or Teo-Alcolhuacan. These nations spoke the same language, they had the same cosmogonic fables, the same propensity for sacerdotal congregations, the same hieroglyphic paintings, the same divisions of time, the same taste for noting and registering everything. The names given by them to the towns built in the country of Anahuac were those of the towns they had abandoned in their ancient country. The civilization on the Mexican table-land was regarded by the inhabitants themselves as the copy of something which had existed elsewhere, as the reflection of the primitive civilization of Aztlan. Where, it may be asked, must be placed that parent land of the colonies of Anahuac, that *officina gentium*, which, during five centuries, sends nations toward the south, who understand each other without difficulty, and recognize each other for relations? Asia, north of Amour, where it is nearest America, is a barbarous country; and, in supposing (which is geographically possible) a migration of southern Asiatics by Japan, Tarakay (Tchoka), the Kurile and the Aleutian isles, from southwest toward the northeast (from 40 to 55 deg. of latitude), how can it be believed that in so long a migration, on a way so easily intercepted, the remembrance of the institutions of the parent country could have been preserved with so much force and clearness! The cosmogonic fables, the pyramidal constructions, the system of the calendar, the animals of the tropics found in the catasterim of days, the convents and congregations of priests, the taste for statistic enumerations, the annals of the empire held in the most scrupulous order, lead us toward oriental Asia; while the lively remembrances of which we have just spoken, and the peculiar physiognomy which Mexican civilization presents in so many other respects, seem to indicate the existence of an empire in the North of America, between the 36th and 42d degree of latitude. We cannot reflect on the military monuments of the United States without recollecting the first country of the civilized nations of Mexico." On a preceding page he also mentions the fact that "the country between the 33d and 41st degrees of latitude, parallel to the mouth of the Arkansas and the Missouri, is considered by the Aztec historians as the ancient dwelling of the civilized nations of Anahuac." The views here expressed, and which all succeeding investigations have tended to verify, carry us back to very remote times, far beyond any authentic history or tradition, when America was peopled by rude tribes of a low grade of humanity, but which, nevertheless, possessing within themselves the germs of a civilization which slowly through the ages evolved a progressive national life, at length resulted in the establishment of the fixed communities in North America, whose skillful husbandry, arts, commercial enterprise and original and complex system of religion we have already contemplated. All of these, however, were but the broad beginning—the prophecy of that higher development which found its fulfillment in the more sumptuous civilization in the rich valleys of Central and South America. The territory occupied by the Mound-builders is too large, the evidences of a dense population throughout its length and breadth too numerous, to permit us to suppose that its occupancy was of short duration. There is also too wide a difference in the respective ages of many of the mounds: some are manifestly hoary with age, while others are of recent date.

While we believe, therefore, that a period of many centuries must

have elapsed during the extension of a people so numerous over the vast
area which they inhabited, and the erection of so many structures as are
still to be seen, it is equally clear from my stand-point that we must also
believe that all facts, when rightly considered, point to a gradual disappearance towards the south, and at different periods of time, which may
be found to correspond to the known dates of the migrations of the Aztec
tribes.   As these occurred at times more or less remote from each other,
it is altogether probable that, to different causes must the separate
migrations be ascribed. Some tribes, as the ruins of their military forts
and encampments show, retreated slowly before the encroachments of an
invading force.   Other sites seem to have been abandoned deliberately,
without any attempt at defence ; or perhaps the impulse which set them
in motion may have been the captivating accounts they had received
of the glory and riches of the distant land to which their brethren had
departed years before. While long lines of military defence may be traced
here and there across the continent at the north, and along the eastern
plains of the valleys of the Mississippi and Ohio, very few have been
observed upon the western side of the Mississippi, at least in Missouri,
and those are of small dimensions. I am led to infer that, however
sudden may have been their abandonment, it was voluntary, and that the
ancient Missourians were the last to leave the country. While it may be
impossible to decide whether they were the Aztecs themselves, or a
remnant of that tribe which was left behind, I cannot forbear to express
my own inclination to the latter opinion ; in which case they may have
proceeded no further than the regions about the Gulf, where they became
amalgamated with the Indians, who may have intercepted them in the
journey, and by whom, as a tribe, they were exterminated.   That some
such event did take place, as before stated, many facts would induce us
to believe. Many of their customs survived them, in the practices of
the more southern tribes, when the country was first occupied by the
Europeans, which point strongly in this direction.   Among these tribes the
Natchez will be remembered, whose arts, worship, sacerdotal system and
customs were very similar, and in many respects identical with those of
Mexico. This identity of customs, worship, etc., I had intended to discuss
more at length, and also present the facts which bear upon the question, but,
as I have already transcended my limits, I must desist.   It seems to me
to be established, that the ancestors of the Indian tribes came to America
by way of Behring's Straits. These are frozen over every year as late as
April, according to Professor Henry, who further states that " intercourse
at present is constant, by means of canoes, in summer between the Asiatic

and American sides. As another fact relating to the same question, we may state that, while the Asiatic projection near Behring's Straits is almost a sterile, rocky waste, the opposite coast presents a much more inviting appearance, abounding in trees and shrubs. Moreover, the climate, when we pass southward of the peninsula of Alaska, is of a genial character, the temperature continuing the same as far down as Oregon. The mildness of the temperature, and the descent of the isothermal line, or that of equal temperature, along the coast, are due to a great current called the Gulf Stream of the Pacific, which carries the warm water of the equator along the eastern coast of Asia, thence across the opposite coast of America, and along the latter on its return to the equator. The action of this current, which does not appear to have been considered by the ethnologist, must have had much influence in inducing and determining the course of the migration." He adds "that the present inhabitants of the countries contiguous to Behring's Straits on the two sides, in manners, customs, and physical appearance, are almost identical." It is believed that the hypothesis we maintain, which holds that the southern portions of this continent were peopled by tribes who had their origin in more northern regions, and who, in some cases at least, were driven from their ancient homes by mongrel hordes who made their appearance by way of Behring's Straits, is the only one which harmonizes the many otherwise inexplicable facts which continually confront the student of the antiquities of America. No other theory will satisfactorily explain the presence in the same mounds of skulls of such different and contrasting types, and which are so frequently met with in the *tumuli* of Missouri.

In bringing our work to a close, I beg leave to say that, in the preparation of the foregoing chapters, it has been my aim to present the subjects treated of in a form as attractive and popular as I was capable of, and in a manner in keeping with the historical character of the work in which they appear. If, to the scientific reader, I may seem at times to have expressed my views with a warmth and enthusiasm not always appropriate to scientific inquiries, my desire to invest with all possible interest, to the general reader, a subject which might ordinarily be considered dry and unattractive, must be my apology. Having for fifteen years devoted all the time which could be spared from the labors of my profession to archæological studies, and especially the antiquities of my native land, the enthusiasm which I felt at the outset has been intensified rather than diminished, at every step of the journey. Indeed, the results which have been attained already are of such absorbing interest as to arouse the

enthusiasm of every student of these antique memorials; and the zeal of
the antiquary receives a fresh impulse from time to time, as he grapples
with those questions which relate to the origin of the different races of
men, their modes of development, the routes of their migrations and the
like; as also, while he labors to construct a pre-historic history from the
ashes of forgotten cities, the debris of former habitations, and the
mouldering relics which ancient tombs disclose.

It is related in sacred story, of an old prophet, that he was set down
in a valley of dry bones, and told to pass by them round about, and
behold they were very many, and very dry. But, at the sound of his
prophetic voice, there was a noise and shaking and coming together, bone
to his bone, the flesh and skin covered them again, and there stood up
an exceeding great army. So the scientist to-day passes up and down
the valleys, and among the relics and bones of vanished peoples, and as
he touches them with the magic wand of scientific induction, these ancient
men stand up on their feet, revivified, rehabilitated, and proclaim with
solemn voice the story of their nameless tribe or race, the cotemporaneous animals, and the physical appearance of the earth during those
pre-historic ages.

The Christian scientist, pursuing his investigations regardless of all
dogmatic theories concerning divine revelation, and bringing, at last, all
right results of his work to the subjective light of that old record which
thus far they have only served to glorify, discovers now and then the
golden key by which the sublime and occult truths condensed in its sen-
tentious statements may be unlocked, and the long æons understood,
which are comprehended in the evening and the morning of the creative
days.

www.ingramcontent.com/pod-product-compliance
Lightning Source LLC
Chambersburg PA
CBHW030613270326
41927CB00007B/1154